HARVEST

A History of Grain Growing, Harvesting and Milling in Ireland from earliest times to the twentieth century.

by

MAJELLA FLYNN

Heritage House
1996

First published in 1996 by ***Heritage House***
Gurteenroe, Macroom, Co. Cork.

ISBN 0 9529707 0 8

Cover illustration: "Threshing", by Joan Jameson

Typesetting, layout, and design: Majella Flynn

Printed by Litho Press, Midleton, Co. Cork

To my mother, Sheila, and my late father, Christy

The publication of this book was made possible by contributions from the following:

Financial assistance toward research work

LEE VALLEY ENTERPRISE BOARD and FÁS

Sponsorship toward publication

National Monuments and Historic Properties Service of the Department of Arts, Culture and the Gaeltacht.

ZENECA Ireland Ltd.

FBD Insurance plc.

IRISH FERTILIZER INDUSTRIES Ltd.

BEAMISH & CRAWFORD plc.
DAIRYGOLD Co-operative Society Ltd.
E. FLAHAVAN & Sons Ltd.
NEVILLE BROTHERS Ltd., Macroom.
TOM CREEDON Ltd., Inchigeela

AIB BANK, Macroom
HENRY GOOD Ltd., Kinsale
HOWARD'S MILL, Crookstown
MACROOM CREDIT UNION Ltd.
MACROOM MILLS Ltd.

DAWN MEATS, Co. Waterford
CANTY FARM SUPPLIES Ltd., Macroom
MACROOM MART
SOUTHERN MILLING Ltd., Cork

Majella Flynn grew up in Toonbridge, Macroom, Co. Cork, and later moved to Gurteenroe, Macroom. She was educated at St. Joseph's Primary School and St. Mary's Secondary School, Macroom, and then at University College Cork where she received her B.A. degree in History and Geography and her M.A. degree in Geography. She has worked in the area of heritage in Co. Cork for a number of years.

Majella is currently studying for a National Diploma in Art at Crawford College of Art and Design, Cork.

TABLE OF CONTENTS

ACKNOWLEDGEMENTS

I wish to gratefully acknowledge the help of the following: the Lee Valley Enterprise Board and FÁS who through the Bealick Mill Restoration Project, Macroom, Co. Cork, provided financial and other support during the research stages of this book; the sponsors who kindly contributed toward printing costs and made publication possible; Denis Lucey, Lee Valley Enterprise Board, who came up with the idea for the book, gave much of his time to proof-reading and commenting on the drafts, and who allowed me access to his collection of catalogues; Martin Fitzgerald, Lee Valley Enterprise Board, who gave his full support to the research initiative, as well as providing office space; Seán O'Sullivan and Brendan O'Brien of the Lee Valley Enterprise Board; the Horgan Family, Firville, Macroom; Pat O'Callaghan, Pat Murphy, Brendan O'Callaghan and Denis Kidney of FÁS; Paddy Healy, Bealick Mill Project; Professor W. J. Smyth, Department of Geography, UCC, who made valuable suggestions for improvements to the text; Dr. Kathleen Curtin who read and commented on the drafts; Breda Duggan and Joe McCarthy of Macroom UDC who kindly gave me access to office facilities; Seán McSweeney, Macroom, for a large amount of the photographic work; Robert Troy, Newtownshandrum, Co. Cork, The Mills Agricultural Museum, Ballyvourney, Co. Cork, and the Lee Valley Open Farm, Kilmichael, Co. Cork, who allowed me to photograph their collections of agricultural implements and machinery; Noel Howard, Donal J. O'Sullivan, Jimmy Moynihan, Jim O'Doherty, Richie Walsh, and the National Museum of Ireland,who provided photographs for publication; Peter Murray, Crawford Municipal Art Gallery, Cork, for help with the front cover reproduction; Denis O'Connell, for his support at various stages of the work; the staff of Boole Library, UCC, Cork County Reference Library, and Cork Archives Institute; and all the researchers whose work I relied upon in drawing together the various strands of grain harvesting and milling in Ireland.

I would also like to thank Marita Foster, Maura McSweeney, Nuala O'Connell, Martin Jolley, Denis Ring, Donal, Teresa and Mary Creedon, Jim Cooney, Cornelius Kiely, and Dan Murphy.

Especial thanks must go to my family and friends who provided continuous encouragement and support through the research, writing, and publication stages of this book.

LIST OF FIGURES

LIST OF PLATES

CHAPTER 1

INTRODUCTION: GRAIN GROWING IN IRELAND'S AGRICULTURAL HISTORY

The wet and mild climate experienced in Ireland is more conducive to grass growing than to grain growing. Consequently, pastoral farming has been a dominant tradition in this country for thousands of years. However, there has been a tendency to overemphasize the importance of this pastoral tradition. Tillage, and grain growing, have been an essential feature of Irish agriculture from earliest times, and have played a major role in Ireland's social and economic development in a number of periods.[1]

The earliest agriculturalists in Ireland were the Neolithic settlers, who reached Ireland before 3000 B.C. These settlers developed a mixed farming economy. While the emphasis was on stock-rearing, they introduced the cultivation of wheat and barley. The arrival of the Celts about 300 B.C. resulted in profound changes in agriculture. More widespread clearance of forest and the cultivation of heavier, more fertile soils were made possible by the use of iron implements. Pastoral farming predominated in this period, but a significant development was the introduction of oats. This crop could be successfully grown on wet, acidic soils and was capable of ripening quickly in the short Irish summer. Barley and wheat continued to be cultivated. In the following centuries, grain growing and milling remained important in the economy. The differentiation of the country into a predominantly arable eastern region and a more pastoral western region began to become evident around the seventh century, as agriculture became more intensive.[2] The arrival of the Anglo-Normans in the twelfth century had a significant impact on agricultural progress. Tillage, and in particular the cultivation of oats and wheat, was important to these settlers, and they were responsible for the further development of the wealthier arable regions. However, there is evidence that some of the land in these regions reverted to pasture land during the fourteenth century. Ireland continued to be predominantly pastoral up to the seventeenth century. However, historians increasingly agree that grain growing continued to be important, and especially in the old medieval arable eastern regions.[3]

The English reconquest of Ireland in the sixteenth and seventeenth centuries had a major impact on Irish agriculture. A strong element of commercialization was introduced and the first half of the seventeenth century witnessed the transition of Ireland from a woodland society to that of one with a considerable agricultural surplus. Tillage continued to be a feature of the Irish-controlled areas during Queen Elizabeth's reign, but, more significantly, there was a noticeable expansion of grain cultivation in the colonized counties. An increase in the population resulted in greater demand for grain products - wheaten bread among the settlers and oatmeal among the native Irish. Along with this domestic demand, a new element of the early seventeenth-century economy began to emerge and this was the export of grain. It was not until the 1690s, however, that the demand for Irish grain abroad became particularly strong. A combination of good harvests in Ireland with harvest failure in Europe, along with problems in the Irish livestock industry which had contributed to the expansion of cereal cultivation, resulted in a grain surplus and the export of this surplus to Scotland and France.[4] The early decades of the eighteenth century were characterized by a succession of boom and crisis harvest years. There were fluctuations in the levels of grain exports. Self-sufficiency in oats, the country's main food cereal, was maintained but by mid-century the country had become a net importer of grain.[5]

From the mid eighteenth century into the early nineteenth century, important economic and social changes took place in Irish agriculture. There was a swing away from pasture into tillage. Grain prices recovered internationally. Britain became more dependent on Irish grain imports, particularly during the Revolutionary and Napoleonic Wars when a blockade of grain supplies from Europe was imposed. Before the turn of the nineteenth century, just over sixteen per cent of Britain's grain imports came from Ireland, but by the mid 1810s this proportion had grown to fifty-seven.[6] Tillage cultivation in Ireland was encouraged by the introduction in 1758 of government bounties on grain and flour destined for the Dublin market. Foster's Corn Law of 1784 increased the export bounties payable on grain, flour and meal.

All of these factors together had a major impact on grain production. At the beginning of the nineteenth century, tillage had expanded dramatically from the core area of the south-east and Ireland had become a net exporter of grain. This product had accounted for about ten per cent of total exports in the 1780s but by 1815 the proportion had increased to about twenty five per cent.[7]

Nineteenth-century Ireland experienced a return to pasture from tillage. However, the precise timing of and degree to which this shift occurred has been the subject of controversy among historians. The ending of the Napoleonic Wars with the Battle of Waterloo in 1815 and the consequent agricultural depression have been regarded as marking a period of major agricultural change. Raymond Crotty, in his 1966 publication, argued against the commonly-held belief that the shift back to pasturage occurred in the post-Famine years. He put a case strongly for a pre-Famine turning point, maintaining that new economic conditions after Waterloo, including falling prices, favoured a return to pasture.[8] However, Crotty's assertions have been weakened by subsequent argument and research. Ó Gráda's analysis of pre-Famine agriculture rules out any significant shift from tillage to pasture. In fact, he points out that tillage items still made up over sixty per cent of total output on the eve of the Famine.[9] As well, grain exports, despite considerable fluctuations in prices, increased dramatically between Waterloo and the Famine. In the mid 1830s, eighty per cent of British grain imports were coming from Ireland, a partial consequence of the protection afforded by the Corn Laws. In response to this strong British demand, tillage acreage increased considerably in pre-Famine Ireland.[10]

While the debate about pre-Famine agricultural trends continues, it is certain that one of the most significant aspects of the post-Famine period was the radical return to pasture from grain and potato cultivation. The growth in livestock products was far greater than that in tillage products.[11] Cattle numbers increased by over sixty per cent and sheep numbers more than doubled during the second half of the nineteenth century.[12] Tillage acreage declined from 4.6 million acres in 1851 to just half that in 1911, with the decline in wheat acreage being particularly noticeable. The already dominant pasture lands of Connaught and Munster experienced the greatest shift into pasture.[13]

The main transition had occurred by 1870. Terms of trade and price trends had an important bearing on this swing to pasture. The Corn Laws were repealed in 1846 and foreign imports to Britain increased. The result was a fall in cereal prices, which displayed a consistency not evident in the pre-Famine period. At the same time, cattle and sheep prices rose. There was an increasing demand for meat and dairy products in Western Europe and Britain after the 1870s, and export of these products from Ireland was made easier with improved transportation. Along with these factors, the decline in population had a significant effect on the tillage/pasture balance. There was a reduction in the demand for arable products and in arable consumption. As well, a declining labour force put a constraint on arable production which was still labour intensive in the post-Famine period. The impact of these factors together was such that by the 1870s - when there was growth in the North Atlantic grain trade and when Britain and Western Europe were experiencing an arable prices depression - Ireland had already adjusted once again to a predominantly pastoral economy.[14]

By the early 1900s, tillage output was only one quarter of total agricultural output and cereal production was in serious decline. There were increases in grain production again during the first half of the century. These occurred around 1919, in the period 1933-36 and in 1944, and were mainly as a response to wartime conditions and tough government measures. While tillage has remained subordinate to cattle throughout the twentieth century, grain production has nonetheless continued to be a significant aspect of Irish agriculture. Though their relative importance has fluctuated considerably over the past half century, barley, wheat and oats are still the three main grain crops grown in Ireland today. [15]

References

1 Smyth, W. J. (1993) 'The Making of Ireland: Agendas and Perspectives in Cultural Geography', 403-4.

2 Gillmor, D. A. (1977) *Agriculture in the Republic of Ireland*, 22-3; Smyth, W. J. op. cit., 404-5; Ó Corráin, D. (1972) *Ireland before the Normans*, 50-1; Simms, A. (1986) "Continuity and Change: Settlement and Society in Medieval Ireland c. 500-1500', 48-9.

References

3 Smyth, W. J. op. cit., 413 and 424; Graham, B. J. (1993) 'The High Middle Ages: c. 1100 to c. 1350', in Graham and Proudfoot, op. cit., 76; Simms, A. op. cit., 64; Gillespie, R. (1984) 'Harvest Crises in Early Seveneenth-Century Ireland', 5-6.

4 Gillmor, D. A. op. cit., 24; Cullen, L. M. (1987) *An Economic History of Ireland since 1660*, 9 and 26-8; Gillespie, R. op. cit., 5-7; Dickson, D. (1987) *New Foundations: Ireland 1660-1800*, 102; Cullen, L. M. (1983) 'Incomes, Social Classes and Economic Growth in Ireland and Scotland, 1600-1900', 249.

5 Cullen, L. M. (1987) op. cit., 43-4 and 67-8; Crotty, R. D. (1966) *Irish Agricultural Production* , 19-20.

6 Ó Gráda, C. (1994) *Ireland: A New Economic History, 1780-1939*, 120.

7 Crotty, R. D. op. cit., 22; Cullen, L. M. (1987) op. cit., 95-6; Whelan, K. (1986) 'The Famine and Post-Famine Adjustment', 153-4; Ó Gráda, C. op. cit., 29.

8 Crotty, R. D. op. cit.

9 Ó Gráda, C. op. cit., 117-8; Ó Grada, C. (1988) *Ireland before and after the Famine* , 50-1.

10 Cullen, L. M. (1987) op. cit., 109; Ó Gráda, C. (1994) op. cit., 120; Ó Gráda, C. (1988) op. cit., 50-1.

11 Ó Gráda, C. (1994) op. cit., 206.

12 Crotty, R. D. op. cit., 84.

13 Turner, M. (1993) 'Rural Economies in Post-Famine Ireland, c. 1850-1914', 294 and 303; Kennedy, L. (1981) 'Regional Specilaization, Railway Development, and Irish Agriculture in the Nineteenth Century', p.183.

14 Gillmor, D. A. op. cit., 26; Turner, (1993)op. cit., 297-8 and 330; Crotty, op. cit., 66; Turner, M. (1987) 'Towards an Agricultural Prices Index for Ireland 1850-1914', 123-4;

15 Turner, M. (1987) op. cit., 126; Gillmor, D. A. op. cit., 26 and 77-80; Crotty, op. cit., 85.

CHAPTER 2

GRAIN GROWING: PREPARATION AND CULTIVATION OF THE SOIL

Grain growing involved a number of stages which spread throughout the farming year. This process began with the preparation and cultivation of the soil, which traditionally took place - weather-permitting - between St. Brigid's Day, February 1st, and May Eve or May 1st. In Ireland, the mild and moist climate has the effect of producing a strong, tough sod which covers the soil. Hence, a necessary first step in most tillage operations in this country was the cutting, lifting and turning of this sod. The spade and the plough were the implements used in this preparation stage. However, while there is evidence for the existence of a beast-drawn plough from early Christian times in Ireland, spade culture predominated well into the latter half of the nineteenth century. Once the sod was turned, the final preparation of the soil before and immediately after sowing was usually carried out with the harrow and the roller. The sowing of the grain seed was traditionally carried out by hand, with mechanization being slowly introduced in the eighteenth and nineteenth centuries.[1]

THE SPADE IN GRAIN CULTIVATION

The spade has been an important and distinctive implement of cultivation in Ireland, since early times. There is evidence that this implement dates back to early Christian times, and possibly to the Bronze Age. In many countries, the spade has been used to stir and shift the soil, but its primary use in Ireland has been to cut and turn the sod. A number of factors have contributed to the predominance of the spade in Irish grain cultivation.[2]

Small-scale farming has been traditional over much of Ireland, and has relied almost totally on simple tools and implements. From early times, the field system known as rundale, or infield/outfield, was a significant element of the agricultural geography of Ireland. This system was in operation over much of Ireland, and particularly in the west and north. It was suited to fragmented areas with scarce arable land but abundant rough grazing, and required few tools or implements. The infield was an area of comparatively dry soil which was broken up into scattered plots, and was around or close to the communal settlement, or 'clachan'. Being under stubble for half the year, wintering cattle grazed and manured this arable land, thus enabling it to produce grain each year. Since the infield was usually kept in permanent cultivation and grain crops were sown only in spring, a light plough was an adequate cultivation implement. However, the infield plots were small and the spade was considered to be a more efficient tool in these conditions. Even when the plough was used, the work was improved upon with the spade. The outfield was located beyond the settlement usually in an uphill area and was used for grazing and periodic cultivation. The spade was used almost exclusively in outfield cultivation, since it was considered the most suitable implement for breaking up the tough sward characteristic of this area. This infield/outfield system continued to be dominant in many parts of the country, and particularly in the west, up to pre-Famine times. While this system went into decline later in the nineteenth century, small-scale farming continued to be important. On this scale, the spade was preferable to the plough, not only because it was cheaper to buy, but also because it was more suitable in small and sometimes stony fields.[3]

It was not until the nineteenth century that the efficiency of spade husbandry as a cultivation technique in both small and large farms became a much-debated topic among agriculturalists. The main criticism of the spade for large-scale cultivation was that while it was efficient in breaking up the soil, its use was slow and labour-intensive and therefore more expensive than horse ploughing. However, it would seem that the value of the spade as a cultivation implement was not questioned by practising farmers. While spade cultivation on large-scale farms may have been more expensive initially than ploughing, this was more than compensated for by better grain yields. In the early nineteenth century, development of effective ploughs was still at a relatively early stage, and it was claimed that the spade dug deeper and tilled more thoroughly.[4] As well as that, there was an abundance of cheap labour in late eighteenth and early nineteenth-century Ireland. Population growth in this period was especially marked among the landless labourers. Many large-scale farmers employed gangs of labourers with spades to till wide fields. Farmers in the east and south of the country often supplemented the local labour supply by hiring migratory agricultural labourers, known as 'spailpíní fánacha'. These labourers, often no more than fifteen years of age, travelled from counties Kerry, Clare, Galway, Mayo, Leitrim and Donegal in spring and autumn.[5] Working with the spade, their numbers were so great in the more prosperous areas of the country that the plough was deemed unnecessary. This labour-intensive spade cultivation was an important contributing factor to high grain yields in Ireland at this time and the continuing dominance of spade tillage in Ireland until well after the Famine was strongly linked to this abundance of cheap labour.[6]

Spade Types, Development and Distribution

The variety of Irish spades used in grain cultivation was enormous, and this was due mainly to differing soil conditions and local custom. Yet, a number of common features may be identified: a rectangular-shaped iron blade with a cutting edge, bent at the centre; a straight or almost straight wooden shaft on which the blade was mounted; a step or foot rest at the top of the blade on which a foot was placed so that the implement could be driven in the ground. One broad distinction can be made between the different spade types, that is, between the one-sided and two-sided spade. The former spade type had one foot rest, on either the left or right of the shaft, while the latter type had two foot rests, one on each side of the shaft.[7]

Until recently, one-sided spades, or loys (láighe), were common in the south and west, while in the north-east, there is no record of their use. In the central area from north Connaught to south Leinster, the one-sided spade has been replaced by the two-sided spade over the past one hundred and fifty years. One-sided spades had long, narrow, iron or steel blades, with open sockets. The shaft, usually straight, fitted either directly into the socket or was wedged in place by a separate piece of wood, which formed the foot rest. This latter construction was common in Munster and south Connaught, while in south-west Ulster and north Connaught the shaft and foot rest were frequently made from a single piece of wood (Figs. 2.1 and 2.2). A spade in an upright position should reach the spadesman's chin, and it was often employed as a unit of measurement in setting out a field for crop-growing. The Munster one-sided spades were distinguished by their long, narrow blades. While there were variations in blade dimensions, two main shapes could be distinguished: those which expanded in a fish-tail form toward the edge of the blade, and those whose widest point was about two-thirds of the way down the blade. The latter tapering blade was suitable for the cultivation of light, gravelly soil. One spade of the fish tail form was the Macroom spring spade, which was sixteen inches long and only three and a half inches wide across the edge. Compared to the Munster blade the Connaught blade was shorter and broader.[8]

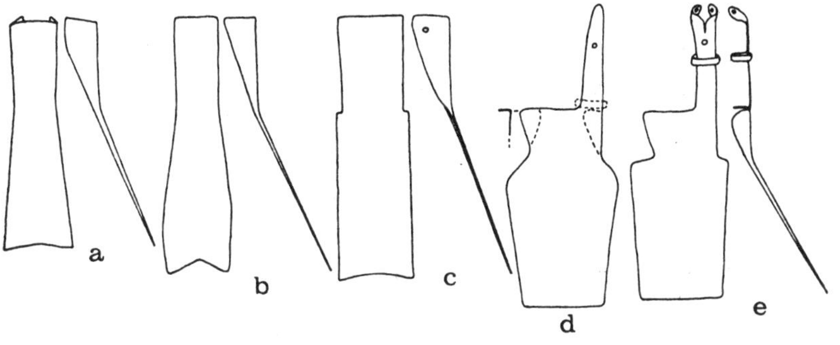

Fig. 2.1 Blades for one-sided spades: (a) Mitchelstown, County Cork; (b) Tulla, County Clare; (c) Swinford, County Mayo; (d) Wexford, County Wexford; (e) Enniscorthy, County Wexford (After Gailey, A., 1970, p. 39).

A peculiar type of one-sided spade was that known as the 'big loy' (Fig. 2.2). Being large and heavy, it could be regarded as having been more a hand-plough than a spade and was found in parts of counties Leitrim and Cavan. It was particularly well-adapted to sod-turning in heavy glacial and peaty soils, and was apparently much quicker than other spade types, especially the two-sided spade. The shaft and foot rest were made from one piece of wood. The former was shaped to have a 'heel' of wood at the back in order to provide extra leverage for the turning of heavy sods.[9]

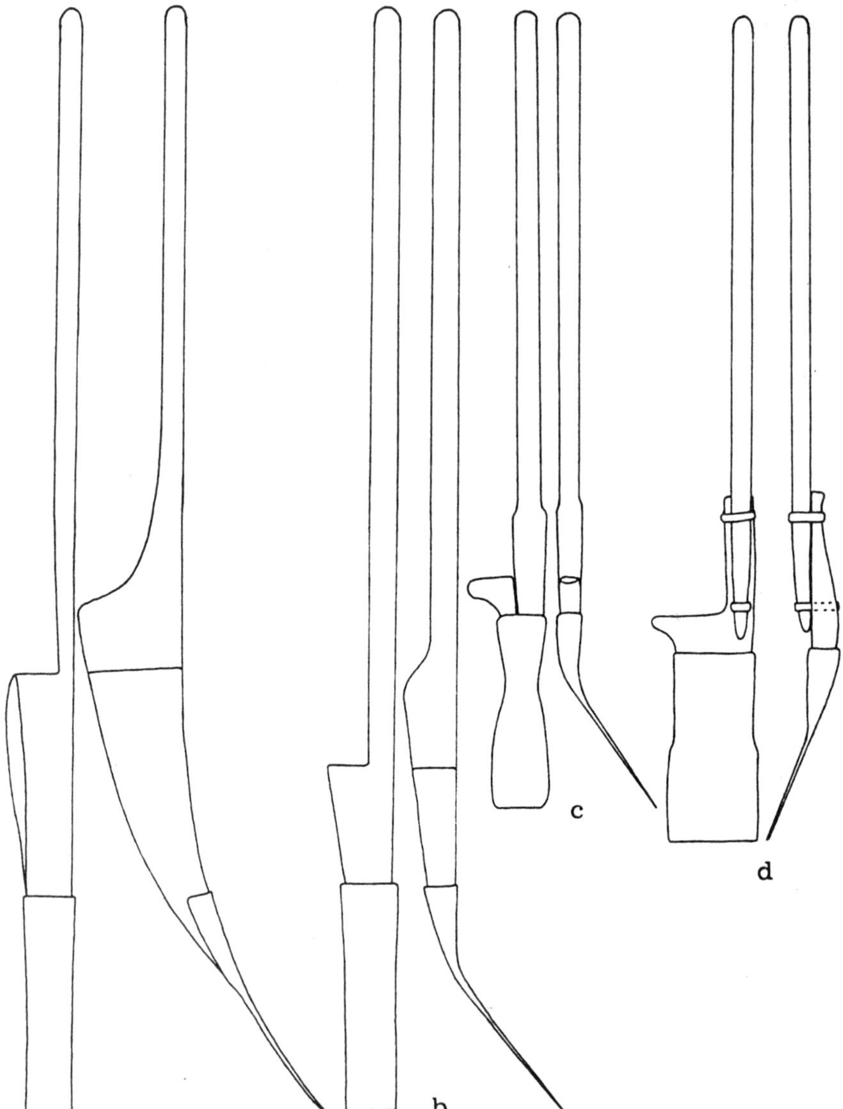

Fig. 2.2 One-sided spades: (a) 'big loy', Mohill, County Leitrim; (b) County Leitrim; (c) Ballinskelligs, County Kerry; (d) The Neale, County Mayo (After Gailey, A., 1970, p. 38).

In many parts of Ireland one-sided spades were replaced by two-sided spades. Originally confined to northern parts of Ireland, two-sided spades increasingly expanded southwards. These latter spade types varied considerably in design (Fig. 2.3). However, generally, the blade is quadrangular, often dished, with straight sides which run parallel or narrow toward the edge. Two-sided spades have a closed socket at the upper end of the iron blade, into which the end of the shaft fits. In the south, west and east, the shafts are generally long, while elsewhere they are short with a timber cross-piece at the top.[10]

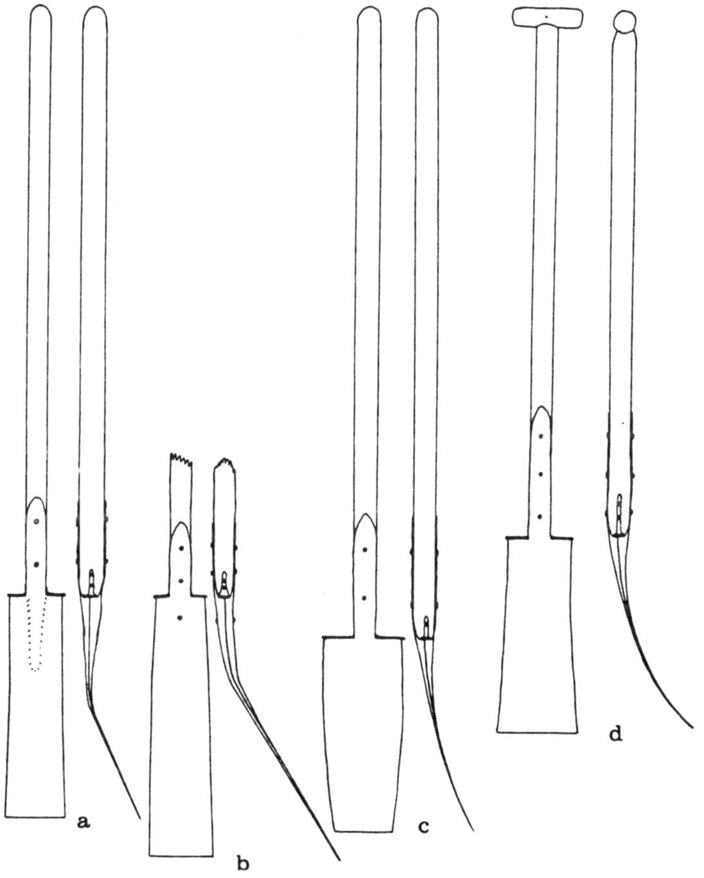

Fig. 2.3 Two-sided spades: (a) Manorhamilton, County Leitrim; (b) Blacklion, County Cavan; (c) Kilkeel, County Down; (d) County Donegal (After Gailey, A., 1970, p. 41).

An area of overlap of the one-sided and two-sided spades covers much of Connaught and extends in a band through the centre of the country to east Leinster (Fig. 2.4). In this area, spade types of a hybrid form can be found. One unique type of spade was the 'gowl-gob' or 'gabhal gob'. This was used in parts of County Mayo but is now extinct. Its distinguishing feature was that it had a two-pronged blade, apparently suited to the cultivation of light, sandy soils (Fig 2.5).[11]

In the south of the country, most spades were made by blacksmiths. Spade-making was a physically demanding task, and involved meeting specific requirements. Factors such as the type and condition of soil of the area in which the spade would be used, type of crop being cultivated, digging methods, and physical needs of the spade user, had to be taken into consideration by the blacksmith. In the north of the country, a different situation pertained, where as early as the mid eighteenth century specialized spade and shovel making mills were being set up. Of the sixty-seven spade mills to have been identified in Ireland, sixty were in the north.[12] The mainly northern distribution of spade mills is explained by the predominance of enclosed socket spades in this part of the country. The manufacture of these spade types was time-consuming, and after the 1780s, blacksmiths were no longer able to cope with the growing demand for such spades. This increasing demand came about as a result of population growth, and because of the growing number of people being forced to cultivate in marginal areas which were not suited to plough use. The blacksmiths of the south did not encounter the same problems, because open socket spade blades, predominant in this area, were manufactured easily.[13] However, in the late eighteenth century and early nineteenth century, a number of spade mills were established in the south. Dublin and Cork were the main centres of large-scale production. In the hinterland of Cork city, five mills had been built by 1837, the largest of which was that built in 1790 at Monard. This latter mill still survives. Standardization in spade types did not occur with the development of spade mills, since the mill-wrights copied the local types in order to compete for custom. The mills merely took over the work of many rural blacksmiths.[14]

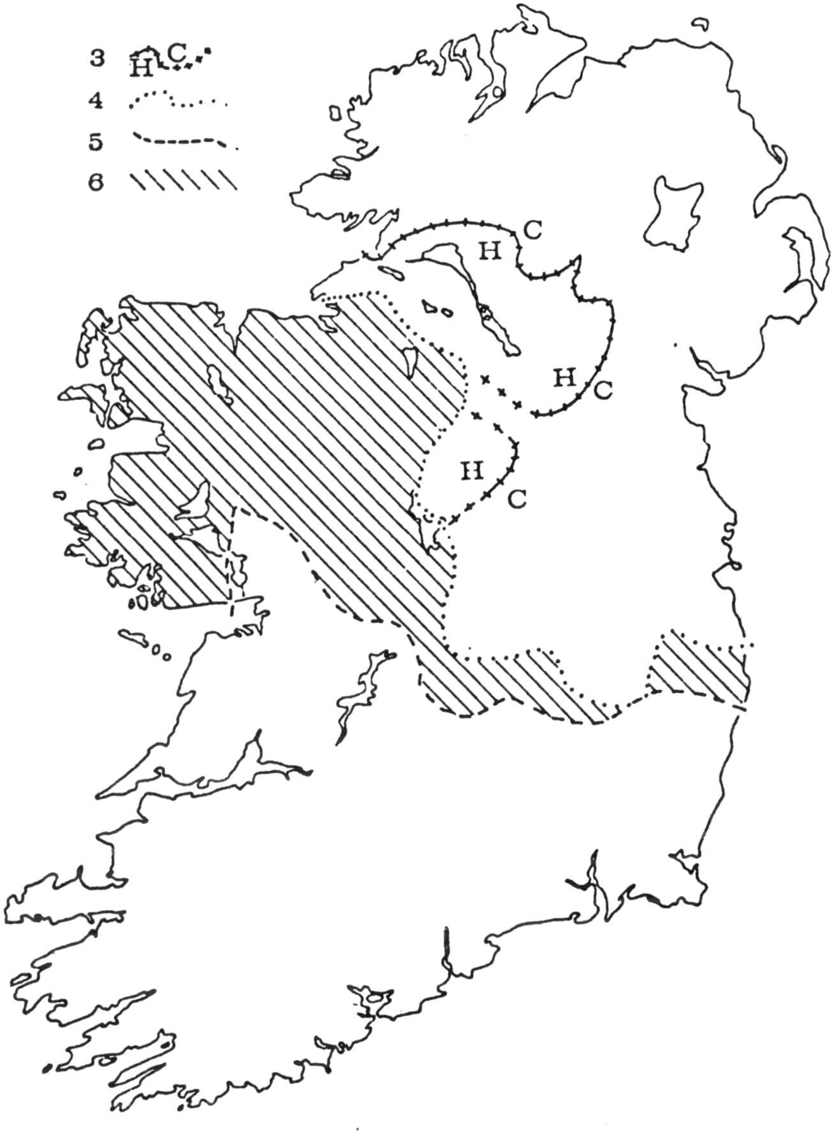

Fig. 2.4 Distributional limits of spade types: 3. boundary between hooked (H) and curved (C) lift in blades of two-sided spades; 4. northern limit of one-sided spades; 5. southern limit of two-sided spades; 6. area of overlap of one-sided and two-sided spades (After Gailey, A., 1970, p. 36).

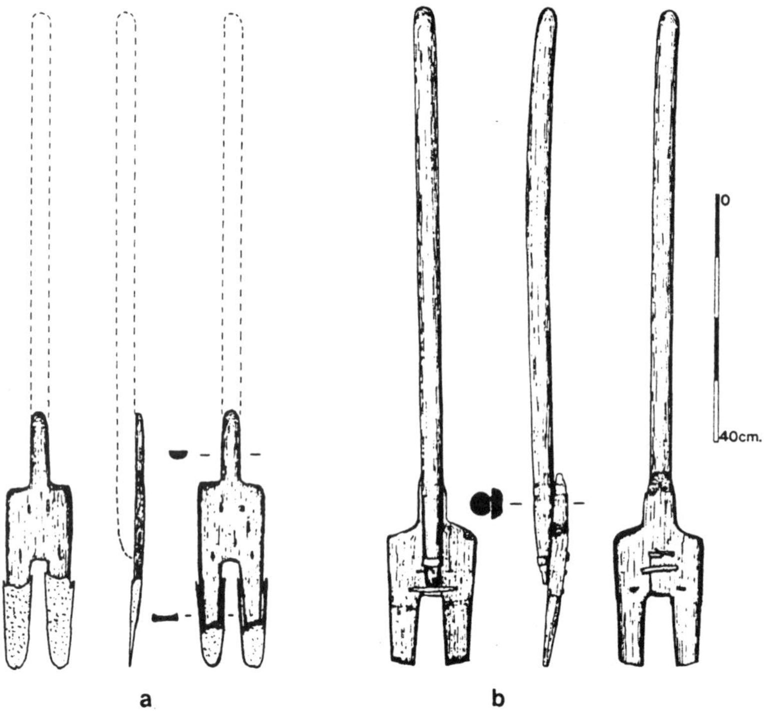

Fig. 2.5 Gowl-gobs (a) from Bartrauve, County Mayo; (b) Locality not recorded (After Lucas, A. T., 1978, p. 195).

Spade Work: Techniques and Practices

Spade work in Ireland was associated particularly with the 'lazy bed' technique, so called by eighteenth and nineteenth century observers who did not understood what they saw. This was a technique of building up ridges on strips of land, which was deliberately left unturned. These ridges were usually only up to three feet wide. Despite its disparaging title, this technique was sensible in the Irish climate, and archaeological evidence from Mayo indicates that raised ridge-making dates back at least to the Bronze Age. Through this practice, drainage was enhanced, the iron-pan in leached soil was broken up and humus and manure were concentrated.[15] The lazy-bed technique was particularly useful in small-scale, labour-intensive farming, in the cultivation of marginal land and of the outfield under the rundale system. Like other aspects of grain growing, co-operation among neighbours was common with ridge-making being undertaken by a team of workers. The first crop to be grown on the 'lazy beds' or 'iomairí' was potatoes. In some areas, for example in west Fermanagh, this crop was sometimes followed by oats, especially at a time when grain was subsequently reaped using hooks. Grain cultivation on high, narrow ridges was common well into the nineteenth century, particularly in the west of Ireland. Some smallholders in the Aran Islands, Donegal and Kerry, still cultivate potatoes in ridges, but the growing of grain on ridges is no longer practiced.[16]

In other parts of the country, a different technique was employed in grain cultivation. If grain was to follow potatoes, or if grain was being sown on recently tilled fallow land on which the sod had not reformed, the whole surface of the field was dug with the spade, and ridges of up to twelve feet wide were constructed. These ridges were more suitable for the mowing of grain with scythes. Grain ridges were often made with the plough, but were shaped and finished with the spade and shovel. Following the growing of a number of crops, the land was left to lie fallow for a period of time.[17]

Mention of the shovel as a grain cultivation implement was made above. The shovel can be described as usually comprising a long shaft, fitted by means of a sleeve socket to a notably dished, pointed steel blade, although up to the Famine, iron-shod wooden spades were common, particularly in the west. The spade and shovel were often used together for the purpose of trenching. This latter practice in Ireland was a part of ridge cultivation, whereby earth was deeply dug from the furrows and spread over the ridge to cover sown grain, potato and flax seeds. The function of the shovel was to lift and spread the broken up soil.[18]

THE PLOUGH IN GRAIN CULTIVATION

Ploughs have been in use in Ireland since the Bronze Age. Under the rundale system, light wooden ploughs were used in the cultivation of grain in the acres of the infield.[19] In the more fertile parts of the country, and especially in the lowlands of Leinster and east Munster where the Anglo-Normans had introduced feudalism, a different field system - the three-field system - was in operation. This involved the subdivision of arable land into three large parts, each of which was used in rotation for autumn grain, usually wheat, then spring grain, usually oats and then fallow. Unlike the rundale system, considerable organization and a greater variety of and sophistication in tools and implements were required to successfully operate this system.[20] It was under this system, and in other fertile areas where large-scale cultivation took place, that the plough was particularly important as a grain cultivation implement. With improvements in plough design in the eighteenth and nineteenth centuries, as well as a reduction in the availability of cheap labour toward the end of the latter century, the plough grew in popularity. By the early twentieth century, the use of the plough was widespread throughout the country, finally supplanting the spade as the predominant implement in grain cultivation.

Plough Types, Development and Distribution

There is evidence that a simple wooden plough type, the ard, was used in Ireland as early as the Bronze Age. This plough type, without coulter or mouldboard, survived in use well into the nineteenth century. It was suitable for light and well-drained soils. Written and archaeological evidence point to the existence of heavy, wheelless ploughs with coulters from the early medieval period. These ploughs produced better furrows and enabled the cultivation of more soil types, including the heavy drumlin clays. A variety of ploughs, considered 'common' by unimpressed agriculturalists, were in use in the eighteenth and nineteenth centuries. Two main types can be identified: the old Irish long-beamed plough and the light Irish plough.[21]

The old Irish long-beamed plough was a heavy, swing or wheelless plough, with a flat wooden mouldboard, a long sole plate and the lack of a plate on the its land side (Fig. 2.6). This type of plough was found especially in the north, east and south of the country. A number of people were required to operate the plough: a ploughman to guide it with its handles, a driver to walk in front and lead the animals, and sometimes a man or boy to lean on it or press it with a stick to keep it steady. A considerable number of horse or oxen, usually between four and six, were required to pull the plough. Surviving descriptive details of the light Irish plough are sparse. There is evidence that it was in use in many parts of Ireland in the early nineteenth century, and was usually drawn by two horses.[22] Its function in ploughing practices is described below.

Fig. 2.6 Old Irish long-beamed plough: Kilkenny plough (Tighe, W., 1802).

From the mid eighteenth century onward, there was growing interest in improving plough design. The improved Scottish swing (wheelless) plough was held up by agriculturalists as being best suited to Irish conditions. The principles upon which this plough was made were laid down by James Small of Berwickshire and later modified by Wilkie of Uddington. The Scottish plough could be operated by one person and required only two horses or oxen. Despite criticisms by agriculturalists of the 'common' Irish ploughs especially in comparison to the Scottish plough, Irish farmers were reluctant to immediately adopt the new plough. The manufacture and operation of the Scottish plough required new skills, its purchase price was relatively high, lower labour requirements implied possible job losses, and it did not always perform well in 'unimproved' cultivation conditions.[23]

However, all-metal Scottish swing ploughs, with the encouragement of 'improving' landlords and farming societies, were increasingly adopted, and by the mid nineteenth century they had become the most popular improved plough type throughout Ireland (Fig. 2.7). Many foundries set up throughout the country began to manufacture the improved plough. These included Paul and Vincent, Le Strange, Sheridan, Kennedy and Graham, all of Dublin, as well as W. & J. Ritchie of Ardee, County Louth and Gray of Belfast. Local blacksmiths began to copy the Scottish plough, and were often supplied with standardized parts by the foundries. The manufacture and construction of the swing plough required much skill and knowledge of ploughing. Wooden ploughs, based on the improved Scottish plough, were also made.[24]

Fig. 2.7 Ransomes' swing plough (Sproule, J., 1854, p. 205).

Swing ploughs continued to be popular into the present century, especially on small, hilly farms. However, the English wheel plough, in fact wheel ploughs of all kinds, became increasingly dominant in Ireland during the later nineteenth century (Fig. 2.8). Large companies, such as Pierce's of Wexford, began to manufacture wheel ploughs based on English design and also on the increasingly popular light American ploughs. The latter had 'digging' mouldboards made of chilled steel, and were suitable for the drill cultivation of grain and other crops. The popularity of wheel ploughs can be related to their lower labour and skill requirements; to their suitability to work on large, well-drained farms, which were becoming more common in the post-Famine period; and to the development of ploughs with interchangeable parts. This latter development proved convenient to farmers, and companies at the forefront included Pierce's of Wexford and the Wexford Engineering Works. In the middle of the present century, the latter firm was making ploughs which could be fitted with parts manufactured in Britain by Ransomes, Howard, Hornsby and Seller, and in America by the firm Oliver.[25]

Ploughing: Techniques and Practices

A wide variety of ploughing techniques, including methods of ridge cultivation, were employed in grain cultivation. Such techniques and practices were dictated by a number of factors, including the type of field system in operation in any one location and the type of ploughs being used. Under the rundale system, ploughing was carried out only in the spring. Winter ploughing was not possible since the inadequate and poor outfield grazing had to

Patent Trussed-Beam Iron Plough, marked Y F L.

This plough is fitted with Ransomes & Sims patent trussed beam; the frame is of wrought-iron, and the mould board is shorter and of a different model to that adopted in our prize ploughs. This plough is extensively used abroad in cases where the land is newly [illegible], in consequence of the great strength of the beam and [illegible] is well suited for resisting strains from [illegible] stones, &c., and being shorter than the [illegible] ploughs, it is more handy in use for such purposes.

RANSOMES AND SIMS' PLOUGH Y F L.

Improved Patent Trussed-Beam Iron Plough, marked Y F R.

This plough is intended for producing furrow s ces of a rectangular section, and of much greater width in proportion to the depth, than is usual in English ploughing.

RANSOMES AND SIMS' PLOUGH Y F R.

Improved Solid-Beam Iron Plough, marked B F S, constructed principally of wrought-iron, and suitable for use with one large or two small horses.

In this plough, which forms one of a series comprising one smaller and two larger sizes, the draught is taken from the body of the plough instead of from the end of the beam, the intention of which is to relieve the beam from strain. In other respects, it is constructed on similar principles to the ploughs previously described, and will make equally good work, but it is only adapted for producing a rectangular furrow.

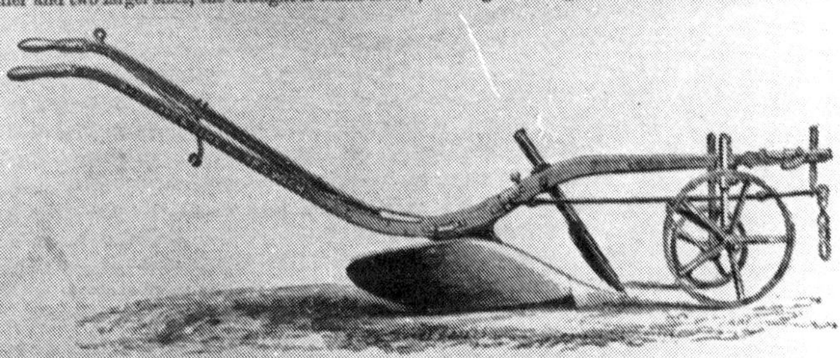

RANSOMES AND SIMS' PLOUGH B F S.

Fig. 2.8 Ransomes and Sims' wheel ploughs (*The International Exhibition of 1862*, Vol. 1, p. 78).

be enhanced with infield stubble. This explains the long absence in Ireland of winter ploughing over many parts of the country. In areas where the three-field system was in operation and where both arable and pasture lands were good, there were no such restrictions, with autumn- and spring-sown grain being grown.[26]

Plough types and ploughing techniques were inextricably linked. The old Irish long-beamed plough was used mainly for the breaking up of ground in winter and spring and for ploughing fallows. The furrow-slices produced were loose and vertical. This plough was used also in some areas up to the early nineteenth century for the formation of curvilinear ridges. Fallowing was a method used in the preparation of land for grain crops. 'Summer fallowing', which preserved land fertility, was widespread in the late eighteenth and early nineteenth centuries. This involved repeated ploughings of the land, especially cross ploughing, during the year. 'Green-fallowing' was employed to break up and cleanse the soil and involved the cultivation of root crops, such as potatoes or turnips. It would seem that the light Irish plough was used where deep ploughing was neither practiced nor necessary, as for example under the rundale system. Some light plough types were used in conjunction with other tillage implements in the primary tillage stages. For example, these ploughs were used in the construction of 'lazy beds' to mark out the ridge by turning the sod on each side. Light ploughs, such as the 'common' plough of County Cork, were used also to effectively cover sown seed with earth. Such plough types were put to similar use in areas with heavy clay soils, for example in the wheat-growing districts of County Kilkenny. A substantial covering of soil was important in the cultivation of grain. The crop was afforded protection from birds and frost, and was allowed to thicken better, resulting in more shoots and a higher yield.[27]

In general, the land was ploughed in narrow ridges varying in width from about three and a half to ten feet. The furrow, about two feet wide, was scraped by the plough to provide soil for covering the grain seeds sown on the ridges. The spread of improved plough types and the extension of land drainage during the nineteenth century brought about standardization in ploughing techniques, particularly on large lowland farms. The construction of broad, low ridges known as 'flats' became common. By the later nineteenth century, ploughing techniques involving ridge construction declined. More widespread land reclamation and other factors such as the adoption of scythes and reaping machines in grain harvesting meant that in many areas ridges were no longer required or desired.[28]

Plough Teams and Tackle

In the early Christian period up to about 1200 A.D., the ox was the usual plough animal, and a team was generally made up of four to six oxen, harnessed, it would seem, side by side. Only the more well-off members of society possessed a full team of oxen and all the ploughing equipment, and so co-operation was necessary. By the sixteenth century, the horse had become the main plough animal, and in this and the following century, draught oxen were confined to the English army and to large estates. In the seventeenth and eighteenth centuries, and in some places into the nineteenth century, a team of four horses pulling abreast was typical. In the late eighteenth, early nineteenth-century period, oxen were again used by large landholders, mainly as part of international experimentation. In the remaining population, oxen were sometimes used when help from neighbours was unavailable and the use of mixed horse and ox plough teams was a practice which was known in the nineteenth and early twentieth centuries. However, by then the use of oxen in Irish ploughing practices was uncommon and the horse continued to be the usual draught animal.[29] During the nineteenth century, with the introduction of the Scottish and English improved ploughs, two horses harnessed abreast became the typical plough team. Small farms could usually support only one horse. Thus, neighbours had to combine their resources to make up a two-horse team in order to undertake the ploughing and later the harvesting work. This co-operative effort was known variously as 'joining', 'cooring', 'morrowing', 'swapping' and 'sharing', and continued well into the twentieth century.[30]

Ploughing tackle comprising ropes made of twisted strips of bog fir, known as withies or 'gaid', was a typically Irish custom going back to ancient times. Trace ropes were made of withies, while the collar, back band, saddle, girths, crupper and bridle were made of plaited straw rope, known as 'súgán'. The use of straw harness continued well into the present century (Pl. 2.1).[31]

Despite the availability of this inexpensive, efficient and humane harness, a notorious Irish ploughing practice was that of 'ploughing by the tail', that is, drawing the plough by attaching it to the tails of the horses. There is written evidence for the existence of this practice as early as 1610 and in 1635 it was prohibited by an Act of Parliament. The practice was considered by its practitioners to be suitable for the cultivation of mountain and stony land, and that the horses, being immediately aware of any obstruction to the ploughshare, would stop at once, thus avoiding damage to the plough and traction ropes. The practice seems to have died out during the eighteenth century, though not completely.[32]

Pl. 2.1 Straw horse collar in use in mid Cork, 1940s (Photograph courtesy of Peter Murphy/Seán McSweeney).

HARROWS, ROLLERS AND THE SOWING OF GRAIN SEED

Harrows and Rollers

The function of harrows was to break up and weed the soil just before sowing, but also to cover the newly sown grain seed by mixing it with the soil. Such tasks could also be carried out with the use of other implements, such as wood mallets (mells), hand hoes or mattocks (graffans, 'grafáin'), grubbers, horse-hoes, and hand rakes. There is evidence that harrows were used in Ireland as early as 700 A.D. and were probably light framed structures with wooden tines (teeth) or simply heavy thorn bushes, known as 'bush harrows'.[33]

From the mid eighteenth century, improvements to the 'common' harrow were being carried out. One successful design was a harrow made in several sections of 'leaves', which were hinged together (Fig. 2.9). The advantages of these hinged harrows were that they lay close to the ground, even on irregular surfaces, and that each section could be easily lifted, useful for the purpose of removing clogged earth. During the nineteenth century, the types of harrows grew, and designs which were generally based on English and Scottish developments became more specialized. Disc harrows, spring harrows, zig-zag harrows and drill harrows became available and were used on the larger farms. Yet, despite these developments, harrows with wooden tines and bush harrows continued in use at least until the end of the century.[34]

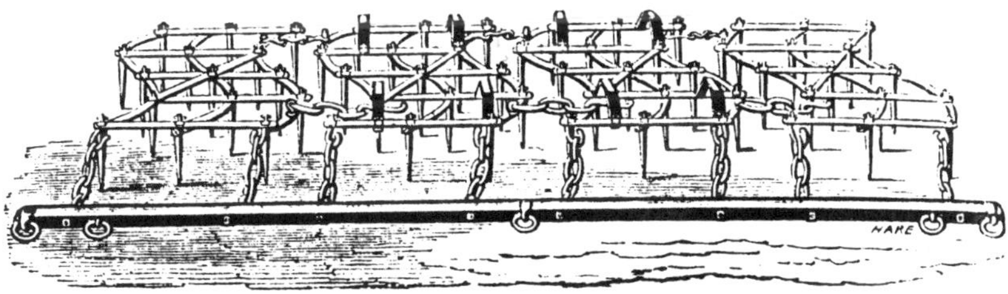

Fig. 2.9 Ransomes and Sims' improved harrow (*The International Exhibition of 1862*, Vol. 1, p. 80).

From earliest times, in contrast to ploughing practices, the most common draught animal used in harrowing was the horse. Horses were adequate in terms of weight, and they were faster than oxen. 'Harrowing by the tail' was practiced up to the nineteenth century, and there is evidence also, from the eleventh century onward, that men and women pulled harrows.[35]

The work of the harrow was completed by the roller. The latter implement was used to flatten and compress the tilled earth before sowing the grain and subsequently to compress the sown seed into the earth. In the last two centuries, rollers were made from wood, stone or metal, and were mounted on a frame, or fitted with shafts either for one or two horses. Rollers made from hollow cast iron cylinders, along with a spiked roller known as Crosskill's clod crusher, were much favoured by agriculturalists and large farmers, in particular (Fig. 2.10). However, on many small farms, mells, rakes, and drag-boards made from pieces of wood nailed to a wooden frame, were used instead of rollers. Traditional rollers, modified for tractor use, are still used on Irish farms, today.[36]

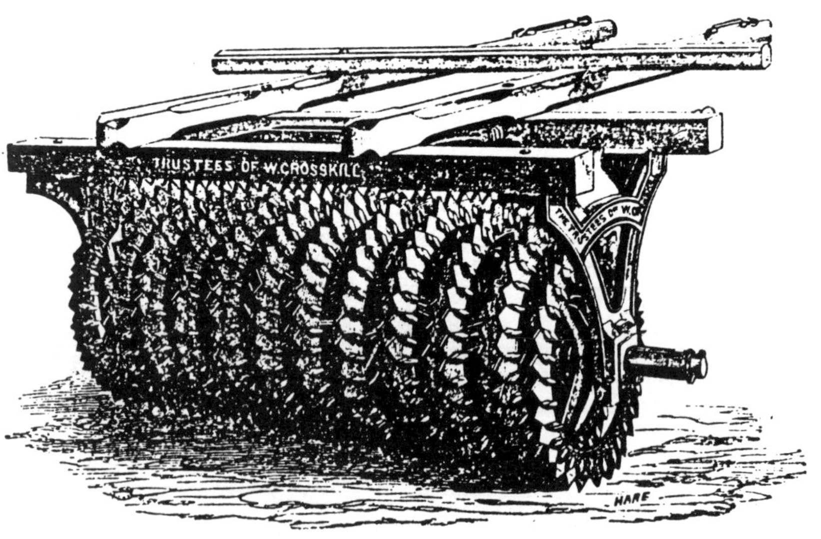

Fig. 2.10 Crosskill's clod crusher (*The International Exhibition of 1862*, Vol. 1, p. 26).

Sowing Grain Seed

Grain seed was planted in either autumn or spring. There were a number of taboos traditionally associated with seed-sowing. In spring-time preparation, Friday, and particularly Good Friday, was considered to be the luckiest day to begin sowing, while no sowing was to be done on Monday. It was considered important that the sower began his work with the phrase "In the name of God" and that he gave the horses some grain seed from his seed bag and threw some earth over their backs. For luck, the ashes of the turf fire were sometimes mixed with the grain seed. Sowing was to be completed by the beginning of May.[37]

From early times, hand broadcasting was the most usual method of sowing, whereby the seed was scattered manually over the prepared ground. The casting of seed in a wide even arc was a very skilled task. While most farmers and farm workers had the ability to carry out this task, there was sometimes one person in the locality who was considered to be an accomplished scatterer. The seed was carried in boxes, baskets, skin trays, purpose-built 'seed lips' and most commonly in pieces of sheet and specially sown cloth bags (Pl. 2.2). This method of sowing continued on smallholdings well into the present century.[38]

Pl. 2.2 A seed lip from which oats was scattered - used in north Cork in the late nineteenth century (Courtesy of Robert Troy).

At the beginning of the nineteenth century, broadcast sowing machines were introduced. Use of these machines allowed a more even spread of seed. One such machine, which had become very popular by the end of the century, was the seed fiddle (Pl. 2.3). This consisted of a bag which held the grain, and a sowing nozzle which opened and closed as the bow was pulled and pushed, in time with the regular strides of the sower.[39]

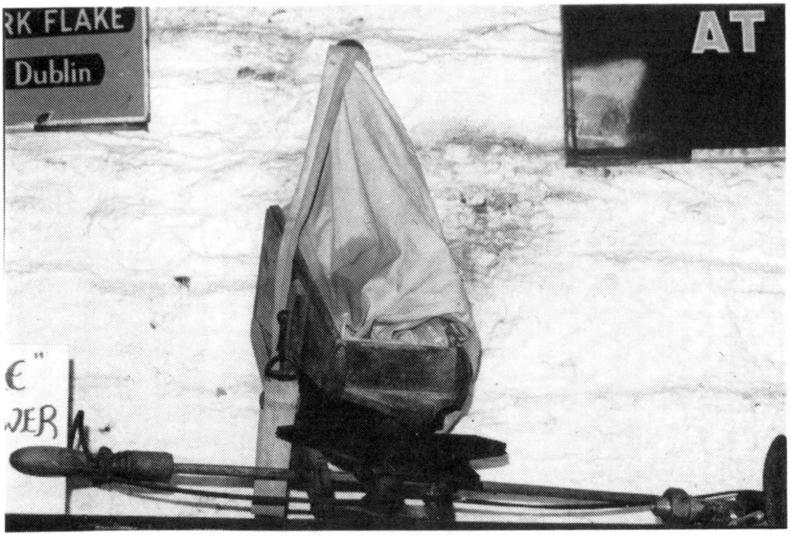

Pl. 2.3 Seed fiddle used in the mid Cork area, *c.* 1930 (Courtesy of The Mills Agricultural Museum).

During the eighteenth and nineteenth centuries, a practice known as dibbing or dibbling seed became common. This involved the sowing, in rows, of seeds in individual holes of a hand-held dibbler.[40] In some areas, four seeds were sown in each hole, and the reason for this is explained in the following rhyme:

"Four seeds in a hole,
one for the rook,
one for the crow,
one to rot
and one to grow"[41]

By planting in rows, hand-hoeing amongst young plants was less difficult. The use of large dibblers for planting grain continued well into the present century in some areas. The main significance of dibbing was that it led the way to a very important development in seed-sowing: the utilization of machines which could sow seeds in rows or drills. Jethro Tull was the most famous inventor of practical seed drills (Fig. 2.11). Horse-drawn and small hand-operated seed drills were developed. The former, which were most significant, usually comprised a large seed box set between the machine's wheels, to which was attached a series of tubes through which seeds fell at regular intervals (Fig. 2.12).[42] Similarly constructed seed drills survived into recent times.

Despite the fact that the task of weeding was made easier by the use of seed drills, the spread of these machines was very slow throughout the eighteenth and nineteenth centuries, and they were generally confined to larger farms in the main tillage regions. Purchase cost and greater draught requirements were among the deterrents to the rapid introduction of these machines. The broadcast method of seed-sowing continued in popularity until the early decades of the present century.[43]

Fig. 2.11 Jethro Tull's Drill, 1703 (After Gahan, W., 1956, p. 252).

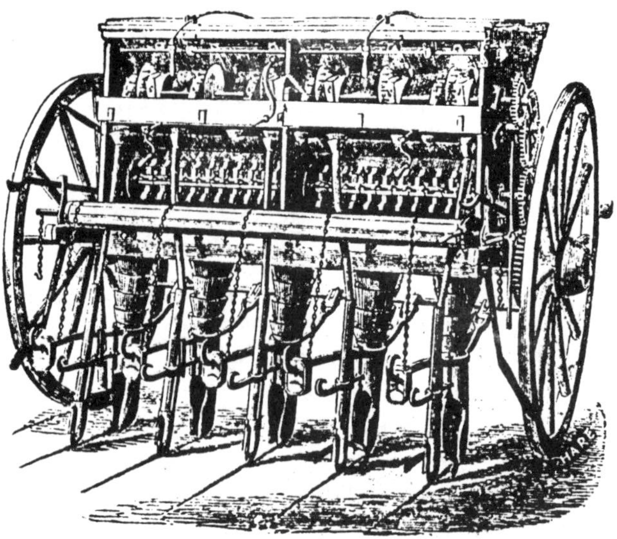

Fig. 2.12 Hornsby and Sons' seed drill (*The International Exhibition of 1862,* Vol. 1, p. 52).

Preparation of the soil for the cultivation of grain was a laborious task, involving the use of the spade, the plough, the harrow and the roller. The predominance of the spade as the most efficient cultivation implement up to the late nineteenth century was influenced by a number of factors: the importance of the rundale field system, small-scale farming and the abundance of cheap labour until after the Famine. The spade, in its variety of designs, was suitable for the cultivation of a wide range of soil types and for the implementation of different grain cultivation techniques. The increasing dominance of the plough from the nineteenth century onward - up to then confined mainly to large-scale cultivation - was to a large degree due to improvements in plough design from the mid eighteenth century and to a sharp decrease in the availability of cheap labour a century later. From the eighteenth century, harrows and to a lesser extent rollers were improved, while the sowing of the grain seed, for centuries carried out by hand, was increasingly done by broadcast machines and eventually by seed drills.

References

1 Pollack, V. (1988) 'Work on the Land', 14-17; Danaher, K. (1962) *In Ireland Long Ago*, 83; Lucas, A. T. (1974) 'Irish Ploughing Practices', (Part 3) 158; O'Dowd, A. (1981) *Meitheal: A Study of Co-operative Labour in RuralIreland*, 70-1.

2 Ó Corráin, D. (1972) *Ireland before the Normans*, 57; O'Sullivan, J. C. (1972) 'Iron-Shod Wooden Spade from Ballynakillew Mountain, Co. Donegal', 246; Bell, J. and Watson, M. (1986) *Irish Farming - Implements and Techniques 1750-1900*, 43; Danaher, K. (1962) op. cit.

3 Evans, E. E. (1957) *Irish Folk Ways*, 8-10, 22-3, 60 and 141; Evans, E. E. (1981) *The Personality of Ireland*, 39-40 and 59-60; Gillmor, D. A. (1977) *Agriculture in the Republic of Ireland*, 23; Whelan, K. (1986) 'The Famine and Post-Famine Adjustment', 154.

4 Bell and Watson, op. cit., 54-5; Ó Danachair, C. (1963) 'The Spade in Ireland', 99-100;

5 Ó Danachair, C. (1970a) 'The Use of the Spade in Ireland', 50; O'Neill, T. P. (1977) *Life and Tradition in Rural Ireland*, 86; Danaher, K. (1962)op. cit., 86.

6 Ó Gráda, C. (1994) *Ireland: A New Economic History, 1780-1939*, 25-6; O'Neill, T. P. (1973) 'Poverty in Ireland 1815-45', 32; O'Neill, T. P. (Undated) 'Tools and Things: Machinery on Irish Farms 1700-1981', 105-7.

7 Danaher, K. (1962) op. cit., 88; Bell and Watson, op. cit., 45.

8 Ó Danachair, C. (1963) op. cit., 101-7 and 113; Gailey, A. (1970) 'The Typology of the Irish Spade', 37; Bell and Watson, op. cit., 45-6; Evans, E. E. (1957) op. cit., 134; O'Neill, T.P. (1977) op. cit.

9 Ó Danachair, C. (1963) op. cit., 108; Evans, E. E. (1957) op. cit., 133-4; Bell and Watson, op. cit., 46 and 54.

10 Gailey, A. (1970) op. cit., 40 and 46; Ó Danachair, C. (1963) op. cit., 111-2.

11 Gailey, A. (1970) op. cit., 36 and 43; Bell and Watson, op. cit., 44.

12 Bell and Watson, op. cit., 43; Evans, E. E. (1957) op. cit., 135.

13 Gailey, A. (1972) 'A family spade-making business in county Tyrone', 26-7.

14 Bielenberg, A. (1991) *Cork's Industrial Revolution 1780-1880*, 93; O'Neill, T. P. (1977) op. cit.; Evans, E. E. (1942) *Irish Heritage*, 116.

15 Gailey, A. (1990) '"...such as pass by us daily...". The Study of Folklife', 14; Ó Danachair, C. (1963) op. cit., 100; Mitchell, F. (1986) 'Early Settlement and Society before 500 AD', 38.

16 Bell and Watson, op. cit., 23-4 and 47; Evans, E. E. (1981) op. cit., 40; Evans, E. E. (1957) op. cit., 141; Bell, J. (1984) 'A Contribution to the Study of Cultivation Ridges in Ireland', 88; Bell, J. (1981) 'Sickles, Hooks and Scythes in Ireland', 32; Mitchell, F. op. cit.

17 Ó Danachair, C. (1963) op. cit.; Bell, J. (1984) op. cit.; Ó Danachair, C. (1970a) op. cit., 54.

18 Gailey, A. (1971) 'Spade Tillage in South-West Ulster and North Connacht', 229; Evans, E. E. (1957) op. cit., 144; Bell and

References

Watson, op. cit., 58; Watson, M. (1985) 'Common Irish Plough Types and Tillage Techniques', 93-4.

19 Bell, J. (1982-83) 'Recent Evidence for the Use of Ard Ploughs in County Fermanagh', 30-1; Evans, E. E. (1981) op. cit., 60.

20 Gillmor, D. A. op. cit.; Evans, E. E. (1957) op. cit., 9; Simms, A. (1986) 'Continuity and Change: Settlement and Society in Medieval Ireland c. 500-1500', 60.

21 Bell, J. (1982-83) op. cit.; Edwards, N. (1990) *The Archaeology of Early Medieval Ireland*, 60-2; Danaher, K. (1962) op. cit., 83-5; Bell and Watson, op. cit., 64.

22 Bell and Watson, op. cit., 65-70.

23 Ibid., 74-5; Watson, M. (1982) 'North Antrim Swing Ploughs',13; Watson, M. (1985) op. cit., 96.

24 Bell and Watson, op. cit., 77-8; O'Neill, T. P. (1977) op. cit., 85.

25 Bell and Watson, op. cit., 79-81.

26 Ibid., 82; Evans, E. E. (1957) op. cit., 141.

27 Bell and Watson, op. cit., 69-70 and 235; Watson, M. (1985) op. cit., 90 and 93-4;

28 Evans, E. E. (1942) op. cit., 90-1; Bell and Watson, op. cit., 26-7 and 85-6.

29 Lucas, A. T. (1972-75) 'Irish Ploughing Practices', (Part 1) 54-60, (Part 2) 67-71, (Part 3) 149, (Part 4) 207-8; Edwards, N. op. cit., 62; Bell, J. (1983a) 'The Use of Oxen on Irish Farms since the Eighteenth Century', 25-6.

30 Bell and Watson, op. cit., 82; Evans, E. E. (1957) op. cit., 23; Pollack, V. op. cit., 14-5.

31 Lucas, A. T. (1972-75) op. cit., (Part 4) 195-6; Bell and Watson, op. cit.

32 Bell and Watson, op. cit.; Lucas, A. T. (1972-75) op. cit., (Part 2) 72 and 78-9; Evans, E. E. (1957) op. cit., 3.

33 Bell and Watson, op. cit., 90; Ó Corráin, D. (1972) op. cit., 52.

34 Bell and Watson, op. cit., 93-6; Bell, J. (1983b) 'Harrows Used in Ireland, 202-3.

35 Lucas, A. T. (1972-75) op. cit., (Part 2) 79; Bell and Watson, op. cit., 97.

36 Pollack, V. op. cit., 14; Bell and Watson, op. cit., 98-102.

37 Evans, E. E. (1957) op. cit., 142.

38 Pollack, V. op. cit., 15; Crowley, F. (1979) *In West Cork Long Ago*, 79; Bell and Watson, op. cit., 103; Evans, E. E. (1942) op. cit., 90.

39 Pollack, V. op. cit.; Bell and Watson, op. cit., 104-5.

40 Bell and Watson, op. cit., 105.

41 Evans, E. E. (1957) op. cit., 147.

42 Bell and Watson, op. cit., 105-7; Evans, E. E. (1957) op. cit., 146.

43 Bell and Watson, op. cit., 108-9; O'Neill, T. P. (Undated) op. cit., 103-4.

CHAPTER 3

GRAIN HARVESTING

The harvesting of grain marked the busiest period in the Irish farming year. Wheat and barley were ready for reaping in August, while the more widespread crop, oats, was harvested in September and early October. A large amount of continuous, hard work was required, work that was over-shadowed by the ever-present threat of rain which could destroy the crops in the field.[1] Migratory agricultural labourers, or 'spailpíní fánacha', played a crucial role in the saving of the grain harvest. Particularly throughout the late eighteenth and early nineteenth centuries, these labourers travelled from the west and south west to the wealthier south-eastern grain growing districts such as Wexford where the demand for reapers in early autumn was significant.[2] Along with the work of paid labourers, the co-operative help of neighbours and friends was essential in order to safely gather in the crop. This band of workers moved from one farm to the next until the grain crops of the locality were saved. Co-operation was particularly important up to the end of the nineteenth century when harvesting was done mainly by hand.[3] While the work was strenuous, the harvest was an important time for festivity. Harvesting of the grain crop involved four stages - reaping, binding, stooking and stacking.

HARVESTING OF THE GRAIN CROP

Reaping

Sickles and hooks were the most common hand-reaping implements used in Ireland, with the larger scythes coming into common use for grain harvesting only toward the later nineteenth century. Sickles and hooks could be distinguished from one another by their blades. The sickle had a saw-toothed blade, whereas the blade of the hook was smooth (Pl. 3.1). The sickle had an advantage over the hook in that it did not require regular sharpening, but reaping grain with the hook was faster. In Ireland, the reaping hook dates back to the Bronze Age. Both archaeological and literary evidence indicates that the hook was the main harvesting implement used up to medieval times, but from the end of this period up to the mid nineteenth century the sickle became dominant. Earlier toothed sickles were made by blacksmiths, but were increasingly replaced by foundry-made implements. The latter were often of an improved design, with the iron teeth being steel-tipped and the blade curved into a semicircle and elbowed above the handle. By the late nineteenth century, these more efficiently designed sickles were very common and were in use even in the remoter western parts of the country.[4]

Pl. 3.1 A reaping hook used in Kilmichael, County Cork, *c.* 1890 (Courtesy of the Lee Valley Open Farm).

Reaping was considered to be one of the most arduous of farming tasks. A variety of techniques were employed using sickles or hooks. A quick method, using the hook, was to slash standing grain which was often held by a fork. However, a neater way was for the reaper to hold a handful of straw in one hand below the grain heads and to cut it toward him/herself with either a hook or a sickle which was held in the other hand. Using this technique, weeds were left behind and very little grain was lost. Reapers usually worked together in a team, or 'boon', along each grain ridge. They were followed by binders who tied the sheaves, which each consisted of about three handfuls or 'loghters' of grain.[5] Reaping with sickles and hooks was carried out by both men and women, while the binding was usually done by women and sometimes children (Fig. 3.1). English evidence indicates that, reaping low, it was possible to cut a quarter of an acre per day using a sickle and a third of an acre using a hook.[6]

Fig. 3.1 A woman reaping with a sickle in the Barony of Forth, County Wexford (*Irish Penny Journal*, 6/2/1841, Vol. 1).

Scythes were used in Ireland as early as the twelfth century on Anglo-Norman manorial estates for the mowing of hay. By the eighteenth century, they had come into common use for this purpose. However, the reaping of grain with scythes was beginning by the early 1800s and became widespread in the latter half of this century. It is likely that this practice was partly introduced by Irish harvest workers returning from Britain. In recent times, a wide variety of scythe types were used in Ireland. These included Scottish Y-shaped shafts, and English and American S-shaped shafts. However, the most prevalent scythe type used in Ireland before the twentieth century had a straight shaft or sned, about five-and-a-half feet in length. Two handgrips were attached to the shaft at right angles and at a distance set for each mower. A piece of wire known as a 'grass-nail', which joined the shaft and blade, was used to adjust the angle of the latter for the cutting of different crops (Pl. 3.2). The scythe required repeated sharpening, and sharpening boards or strickles and scythe stones were both widely used for this purpose. Sometimes, an iron spike was attached to the end of the shaft to secure the scythe safely to the ground when sharpening. The scythe 'cradle', an attachment for catching the swathe as it was cut, became common during the nineteenth century. Many of these cradles were makeshift but more elaborate ones made of wood and wrought iron were to be found in the south-east where such attachments seem to have been first used.[7]

Cutting grain with a scythe required skill. Either of two methods was employed by the scythesman. Cutting along the edge of the crop, the blade was swung either toward or away from the grain, known respectively as 'cutting in' or 'cutting out'. Easier cutting was made possible by the technique of 'rodding' or 'poling', whereby the grain to be cut was pressed against the main body of the crop with a rake or pole usually held by a young boy or girl. Once cut, the sheaves were tied by another person. Between one and two acres per day could be cut using a scythe, with 'cutting in' being the faster method.[8]

The scythe increasingly replaced the sickle and hook as a grain harvesting implement from the second half of the nineteenth century. However, there were a number of reasons for its relatively late introduction, in Ireland. The use of high, narrow cultivation ridges continued well into the nineteenth century, as did broadcast sowing techniques which made weeding difficult. Additionally, the use of the spade meant that surfaces were uneven. These techniques militated against the use of the scythe, which was most suited to level ground and for the cutting of crops that were evenly sown and free from weeds. It was only from the middle of the nineteenth century when cultivation ridges were flattened with the onset of large-scale drainage schemes, and when more systematic weeding was being developed, that the scythe could be used more efficiently. Mowing with a scythe was more than twice as fast as reaping with a sickle or hook. Except in large-scale farming, however, this was not a great advantage since labour was abundant, especially up to the latter half of the nineteenth century.[9]

Pl. 3.2 A long-handled scythe in use, mid Cork (Courtesy of Seán McSweeney/ Lee Valley Open Farm).

From the late eighteenth century, attempts to develop reaping machines were being made in England and Scotland. The most practical of the early machines was that developed in the late

1820s by the Scot, Patrick Bell. However, only a few of this design were manufactured and used (Fig. 3.2). It was in America that the most important advances were made. Two American reaping machines, McCormick's and Hussey's, were shown at the Great Exhibition in London in 1851, and generated much enthusiasm (Fig. 3.3). These prototypes were successfully developed and improved by manufacturers. Unlike the earlier inventions, including Bell's reaper, these American machines were pulled rather than pushed by horses. By the late 1850s, combined reaping and mowing machines were being developed, their large wheels providing the power for the cutting knife of the reaper. From the 1860s onward, reaping machines manufactured in Britain and America, as well as in Ireland, were available on the Irish market (Pl. 3.3). Pierce's of Wexford was the largest Irish manufacturer. The Irish machines were usually based on well-known designs. In comparison to hand implements, the cost of purchasing a reaping machine was very high. However, the shorter harvesting time and lower labour requirements led to savings, but mainly in large-scale farming. Reaping machines were popular in the principal tillage areas by the end of the nineteenth century, but sickles and hooks continued to be used successfully on the smallest farms.[10]

Fig. 3.2 An 'improved' version of Bell's reaping machine (*The Illustrated London News*, 15/11/1862).

Fig. 3.3 McCormick's reaping machine (*The Illustrated London News*, 19/7/1851).

Pl. 3.3 'Albion' - a one horse manual reaper manufactured by Harrison McGregor & Cos. (Courtesy of Robert Troy).

The introduction of reaping machines was followed by further technological advances in grain harvesting. In 1887, McCormick's began marketing reaper-binders. These new machines reaped but also tied the straw into sheaves with string. While harvesting costs per acre were reduced, these machines were expensive to buy. It was well into the twentieth century in Ireland before these machines were purchased even by large farmers. Gradually, however, reapers and reaper-binders began to replace sickles, hooks and scythes (Pls. 3.4 and 3.5). These machines were sturdy and long-lasting, and they continued to be used until many of them, in turn, were replaced by the combine harvester later in the twentieth century .[11]

Pl. 3.4 A reaper and binder harvesting grain in mid Cork in the 1950s. In the background, the tied sheaves have been put in stooks by the workers (Photograph courtesy of J. P. Kelleher/Seán McSweeney).

Pl. 3.5 A reaper and binder harvesting grain in mid Cork in 1947. The standing grain, the packed sheaves which were thown by the machine every ten to twelve feet, and the stooks, can be seen (Photograph courtesy of Denis Lucey).

Binding, Stooking and Stacking

Before the introduction of binding machines, the binding of grain sheaves was a skilled and speedy task usually done by women. There was sometimes just one binder to two or more reapers. The sheaf - with the heads of grain all in the one direction - was laid out on a band made of straw, with which it was then tied. In many areas of the country, a strong band was made by splitting a handful of grain at the stubble end and twisting the stalks around one other. This was known as a 'double band'. It was important that the binding was secure, but not so tight as to prevent the circulation of air. In order that the sheaves would stand properly in the stooks in which they were subsequently put, the bands were not placed too close to the stubble ends, but usually around the middle of the sheaf.[12]

Stooks varied in construction and in the number of sheaves they contained. One type of stook was constructed by a method known as 'pirling', whereby four sheaves were set up and tied together at the top. The more common types of stooks were made up of ten or twelve sheaves standing in two rows, with two 'hooding' or 'head' sheaves laid on top for protection against rain and birds. These stooks had an initial construction stage of footings of four to six sheaves. After one or two weeks of seasoning, these stooks were, in some areas, built into small, round hand-stacks, variously known as 'barts', 'huts', 'shigs' and 'adags'. When dry, the stooks or the stacks into which they had been made were transported from the fields by various methods - on the back using a burden rope, by rick-shifter or cart, or later by tractor lifter.[13]

The drawing in of all the grain and the subsequent construction of large stacks in which it was stored required a large work force. These stacks were built in the 'haggard' or 'stack-yard' on larger farms and in the corner of a yard or field on smaller farms (Pl. 3.6). In Ireland, stacks were usually built on circular bases. Some of these bases were simply made of branches of furze or thorn laid on the ground, but 'corn-stands' which raised the stacks off the ground and deterred vermin attack were also used (Pl. 3.7). Various types of stands were built. These included crude bases made from local, flat stone, more elaborate stone bases, as well as metal stands. Care was required in the construction of stacks. This job was usually carried out by a skilled 'stack-maker'. It involved placing the sheaves in layers, with the heads pointing inwards and meeting at the centre. It was important that the sheaves sloped slightly upwards toward the centre so that rain water would drain off. Long-handled light forks or 'pikes' were often used to throw the last sheaves to the top. The completed stack tops were cone-shaped. The stacks were frequently thatched with rushes and tied down with ropes for protection against rain and wind, respectively (Fig. 3.4). The grain was stored in these stacks until such time as threshing took place.[14]

HARVEST FESTIVITIES AND CUSTOMS

Co-operation was essential in the harvesting of the grain crop. The importance of co-operative labour is highlighted by the large variety of terms used in Ireland to describe it. 'Meitheal' was the most common term, its usage being recorded in twenty-two of the thirty-two counties of Ireland. In relation to harvest work, it was

Pl. 3.6 A 'stack-maker' constructing a stack of oats near Belmont, County Offaly, in 1969 (Photograph courtesy of the National Museum of Ireland).

Pl. 3.7 Corn stacks on stone stands near Athenry, County Galway, in 1970 (Photograph courtesy of the National Museum of Ireland).

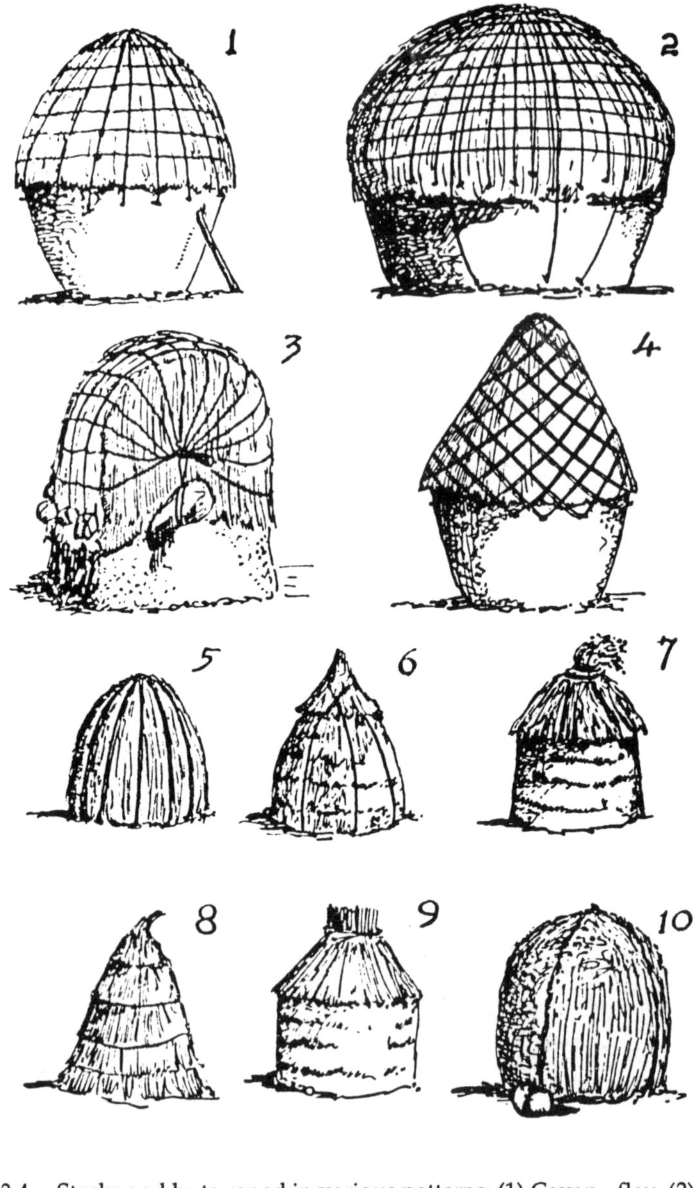

Fig. 3.4 Stacks and huts roped in various patterns: (1) Cavan - flax; (2) Cork - hay; (3) Mayo - hay; (4) Antrim - oats; (5) Donegal - hay; (6) Donegal - oats; (7) Down - oats; (8) Down - seed hay; (9) Cavan - oats; (10) Down - trampcock of hay (After Evans, E. E., 1957, p. 154).

usually used to describe a band of reapers.[15] Another term was 'boon', confined to the northern counties, and referred to in the 1866 poem by Flecher, entitled *The Churn*:

> "A boone attacks a golden field
> in bright September morn,
> And keen the glancing hooks they wield,
> And fast they fell the corn."[16]

Large gatherings of workers, paid and unpaid, were thus a central element of the harvest period. While the work was strenuous, in part brought about by rivalry and competition among the workers, harvest time was characterized by gaiety and was a very important social occasion for men and women.[17] Before the spread of mechanization which speeded up the whole harvesting operation and reduced labour requirements, the harvest was strongly associated with festivity, custom and ceremony.

The Festival of Lughnasa

In Celtic times, the year was marked by four festivals: Samhain on November 1st, Imbolc on February 1st, Beltaine on May 1st and Lughnasa on August 1st. The latter festival was a celebration of the beginning of the harvest of the main subsistence crops: potatoes in the last two centuries and grain in earlier times. This festival has survived down to the present day under a variety of names, including 'Lammas Sunday', 'Garland Sunday', 'Height Sunday', and 'Bilberry Sunday', and takes place on the last Sunday of July or the first Sunday of August.[18]

Lughnasa was the traditional date for the first cutting of the grain crop or the first digging of the potato crop, and it was considered inappropriate to do so before this date. This day was also marked by the meal of first fruits of the harvest. While this meal has been associated with the potato in more recent times, the memory of the festive meal made from grain survives. It was traditional for the new grain to be reaped in the morning and served as porridge or bread before evening. This custom necessitated the speeding up of the preparation of grain for food. This was done by a process known as 'burning in the straw', whereby the sheaf was burned to separate the grain from the straw and to eliminate the chaff and harden the grain. The latter was then ground in a quern, and cooked or baked. Fresh fruit was also served at the festive meal.[19]

A very important aspect of the festival, and even more so than the festive meal in many areas of Ireland, was the festive gatherings which took place at traditional outdoor sites. Mountain or hill tops, lakes, rivers and/or wells were the traditional locations for such festivities. Until recently, the mountain and hill gatherings numbered seventy-eight, and were spread throughout Ireland. Up to fifteen river or lake gatherings took place, mainly in the north midlands. The festivities at these sites involved eating, drinking, singing, games, dancing and courting. The picking of wild fruit, especially bilberries, and wild flowers were important aspects of the festival. Weather divination, which was meaningful in the context of the harvest, was also a common element. A further seventeen height gatherings of secular origin have been converted to places of pilgrimage, the most famous being Croagh Patrick. There were many instances of excursions to holy wells and shrines at this time of the year, though a lot of these may not be directly connected with the harvest festival. The holding of fairs was common, and again while many have more recent origins, some are survivals of Lughnasa, for example, Puck Fair in Killorglin, County Kerry (Fig. 3.5).[20]

Saint Bartholomew's Day

By St. Bartholomew's Day, on August 24th, the harvest should be well in progress. On this day, the flails were prepared for the threshing of the grain. A high wind at this time was said to be 'Beairtlí na Gaoithe' ('Bartholomew of the Wind') testing his flail or preparing it to thresh the grain still standing in the field.[21]

The Last Sheaf and Harvest Feast

Throughout Ireland, there was much ceremony attached to the cutting of the last sheaf remaining in the field. This last sheaf was known by a variety of names, such as 'cailleach' and 'hare'. Standing a short distance from the sheaf, which was often plaited, the reapers threw their sickles or hooks at it in turn until it was cut. The person who succeeded in cutting the last sheaf was loudly cheered, though in some parts of Ireland it was believed that bad luck would befall the cutter. In some places, scythes were used but their unsuitability for the task combined with their increasingly

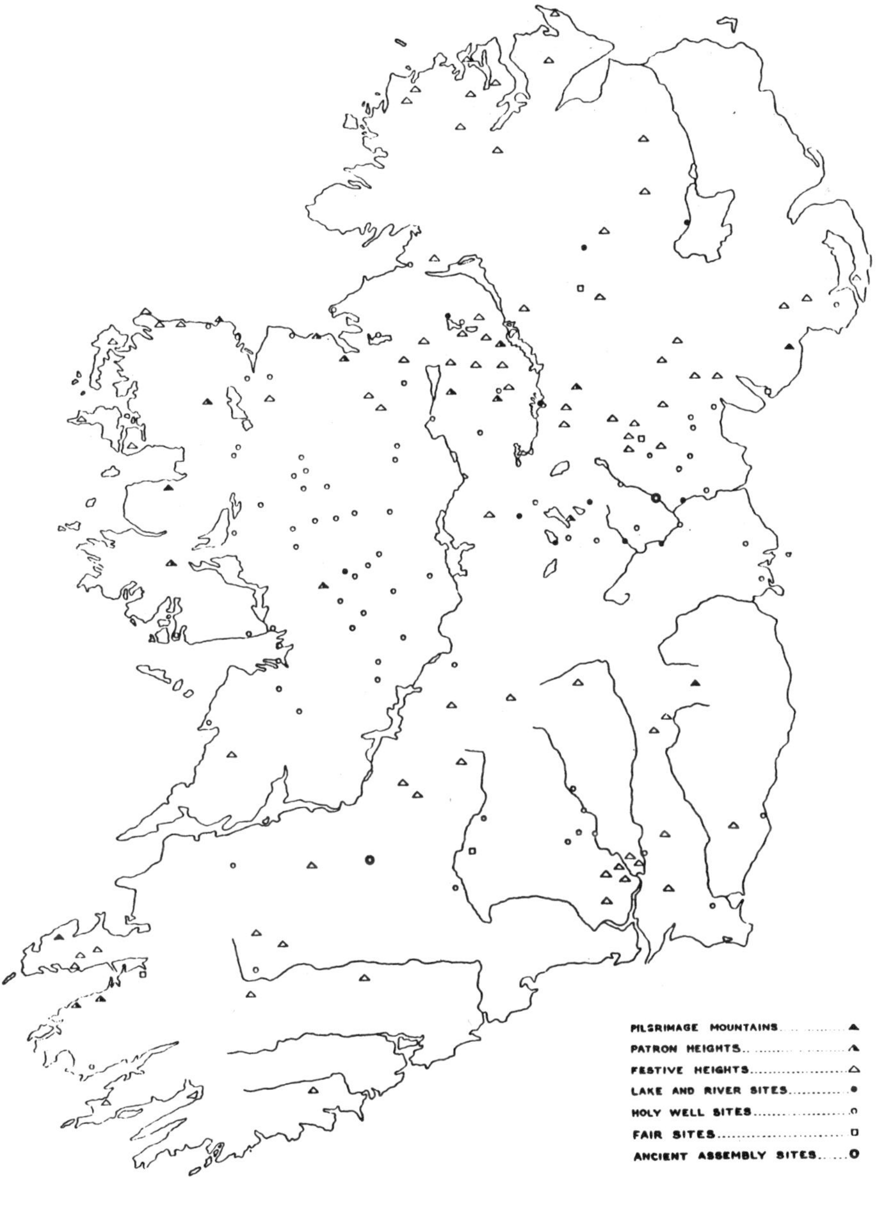

Fig. 3.5 Sites of outdoor gatherings associated with the Festival of Lughnasa (After MacNeill, M., 1982).

widespread use as a harvesting implement contributed to the decline of the custom. The last sheaf was carried from the field most usually by the cutter, who was accompanied by the other workers. At the farmhouse door or in the farmyard it was presented to the farmer, who was expected to provide drinks or a harvest feast in return. This feast, known by such names as 'harvest home', 'churn', and 'clabhsur', was held in the farmhouse or grain barn. Usually, the last sheaf was displayed at the feast and the subsequent dance, after which it was kept until the next harvest.[22]

'Burying the Sheaf'

A more sinister custom associated with the grain harvest, especially in the east midlands and in north Antrim, was what was known as 'burying the sheaf'. A sheaf, plaited into a human or heart shape, was buried underground and was cursed in the name of an enemy. The belief was that as the grain decayed the named person died also. It seems that the only way that this person could avoid such a fate was to locate and unearth the sheaf and then dry or burn it.[23]

Harvest Knots

A harvest custom which was most common in Ulster, east Cork, eastern Offaly and Laois and north County Dublin was the making of harvest knots (Fig. 3.6). These were of two types: a decorative twist of two straws with the grain ears attached and a less elaborate plait with the ears trimmed off. The former type was worn by women in their hair, while the latter was displayed by men on the lapel of a coat or on a cap. These knots were worn throughout the harvest time and were often exchanged as tokens of love. In some cases, they were specially made for and worn at the harvest feast or dance. The making of harvest knots survived into the twentieth century.[24]

Until the second half of the nineteenth century, the cutting of grain was carried out with sickles and hooks, and from this period by scythes. The introduction of the latter had been delayed by the use of high, narrow cultivation ridges, broadcast sowing techniques and the use of the spade. Reapers and reaper-binders, initially confined to the main tillage areas and to large-scale farms, gradually replaced the hand tools, and all were replaced by the combine harvester during the twentieth century. Before mechanization, binding was a skilled job usually done by women, and the stooking and stacking that followed also demanded much labour. The saving of the harvest was marked by much celebration and this time of the farming year was punctuated by festivity, custom and ceremony.

Fig. 3.6 Harvest knot from County Armagh (After Danaher, K., 1972, p. 193).

References

1 Evans, E. E. (1957) *Irish Folk Ways*, 151; Edwards, N. (1990) *The Archaoelogy of Early Medieval Ireland*, 62.

2 Johnson, J. H.(1967) 'Harvest Migration from Nineteenth-Century Ireland', 103.

3 Buchanan, R. H. (1988) 'Calender Customs', 53; Bell, J. and Watson, M. (1986) *Irish Farming - Implements and Techniques*, 181.

4 Bell and Watson, op. cit., 181-2; Bell, J. (1981) 'Sickles, Hooks and Scythes in Ireland', 27.

5 O'Dowd, A. (1981) *Meitheal*, 74; Bell and Watson, op. cit., 182-3; Evans, E. E. (1957) op. cit., 161; Pollack, V. (1988) 'Work on the Land', 19; Gailey, A. (1990) "'...such as pass by us daily...': The Study of Folklife',10.

6 Bell and Watson, op. cit., 185 and 189; Bell, J. (1981) op. cit., 29; Evans, E. E. (1942) *Irish Heritage*, 101.

7 Bell and Watson, op. cit., 185-8; Bell, J. (1981) op. cit., 30-1; Evans, E. E. (1957) op. cit., 155-7; Sharkey, O. (1985) *Old Days, Old Ways*, 119-120.

8 O'Mahony, S. *A Gate to the Past*, 165; Bell and Watson, op. cit., 189.

9 Bell, J. ' (1984) A Contribution to the Study of Cultivation Ridges in Ireland', 94; Bell and Watson, op. cit., 189-192; Bell, J. (1981) op. cit., 33; Ó Gráda, C. (1994) *Ireland: A New Economic History, 1780-1939*, 92.

10 Bell and Watson, op. cit., 192-4; Blandford, P. W. (1976) *Old Farm Tools and Machinery*, 119-120; Gahan, W. (1956) 'The Origins of Crop Husbandry', 253; O'Neill, T. P. (Undated) 'Tools and Things: Machinery on Irish Farms 1700-1981', 109.

11 Bell and Watson, op. cit., 195-6; O'Neill, T. P. (1977) *Life and Tradition in Rural Ireland*, 87; Blandford, P. W. op. cit., 124-5.

12 Pollack, V. op. cit., 19; Evans, E. E. (1957) op. cit., 161-2; Bell and Watson, op. cit., 196-7; Lucey, D. pers. comm.

13 Evans, E. E. (1942) op. cit., 101-2; Evans, E. E. (1957) op. cit., 164; Bell and Watson, op. cit., 198-9; Pollack, V. op. cit.; Sharkey, O, op. cit., 124-5; Lucey, D. pers. comm.

14 O'Dowd, A. op. cit.; Bell and Watson, op. cit., 201-2; Evans, E. E. (1942) op. cit.

15 O'Dowd, A. op. cit., 36 and 57.

16 Ibid, 38.

17 Ibid, 74; Gailey, A. op. cit.; Buchanan, R. H. op. cit.

18 MacNeill, M. (1982) *The Festival of Lughnasa*, vii and 1; Evans, E.E. (1957) op. cit., 275.

19 MacNeill, M. op. cit., 49 and 60; Danaher, K. (1972) *The Year in Ireland*, 167-8.

20 MacNeill, M. op. cit., 68, 289, 427; Danaher, K. (1972) op. cit., 170-2 and 175-6; Buchanan, R. H. op. cit., 51-2.

21 Danaher, K. op. cit., 179; Evans, E. E. (1957) op. cit., 276.

22 Danaher, K. op. cit., 191-3 and 198; Evans, E. E. (1957) op. cit., 163; Gailey, A. op. cit., 11; O'Dowd, A. op. cit., 114-5.

23 Conboy, N. (1982-83) 'Burying the Sheaf ', 102-4.

References

24 Danaher, K. op. cit., 196-7 and 198; Evans, E. E. (1942) op. cit., 177; Buchanan, R. H. op. cit., 56; Paterson, T. G. F. (1944) 'Harvest Customs in Co. Armagh',110.

CHAPTER 4

THRESHING, WINNOWING AND STORAGE OF GRAIN

In the period following the harvest, the threshing of the grain - that is, the separation of the seed from the straw - was undertaken. Up to the early nineteenth century, threshing was done almost exclusively by hand, with each farming household possessing the tools necessary to carry out this task. Threshing of the stored grain stacks was usually spread over the winter and spring months, particularly when the grain was required for domestic use only. The introduction of mechanization, and especially steam threshing, speeded up the whole process. During the month of October, these machines moved from farm to farm, threshing each crop in a few days. A large work force, both voluntary and paid, was required to keep up with the speed of the thresher while it was at the farm. Consequently, threshing became a big occasion and was a social highlight in the farming year.[1] The steam thresher also winnowed the grain, that is, separated the husks or shells from the threshed grain. Prior to the introduction of threshers which could also winnow and on farms which did not use such machines, this process was carried out by hand or by winnowing machine.[2] The threshed grain was either sent to market or stored on the farm for home use.

GRAIN THRESHING

Traditional Methods of Threshing

A number of methods were traditionally used in Ireland to thresh grain. These included 'burning in the straw', 'lashing', stick threshing and flail threshing. 'Burning in the straw' was carried out in either of three ways. The whole sheaf was burned and the parched seed was sifted from the ashes; the ears were set alight and the grain with the chaff burned away was knocked off with a stick; or the ears were cut off from the sheaf and burned in a heap. It seems that this method was used when a small quantity of grain was needed to make bread or porridge. Burning prepared the grain for grinding, achieving in one quick operation the combined results of threshing, winnowing and drying. While this method was forbidden by an act passed in the Dublin parliament in 1634, memories of bread made from burned grain, known as 'loiscreán', have survived until recently .[3]

Lashing was a simple operation. The sheaf was grasped in both hands at the stubble end and the ear end was beaten against a hard object placed at a suitable height, such as a large stone, a chair, a barrel, or a ladder. In this way, the grain was dislodged from the ears. As it was not crushed it was suitable as seed for the crop of the following year. Straw which was lashed was favoured for thatching since it was left unbroken. This method was used in many places to acquire small quantities of grain before the bulk of the crop was threshed by other means, while it was the only technique used in some mountainous areas of Kerry, Galway and Mayo where crops were small. Lashing is still occasionally carried out if only a very small amount of seed is required.[4]

Stick threshing, known as 'scutching' or 'scoth-bhualadh', was another method employed for similar purposes. Usually, the sheaf was held in the hand or placed on a stool or box and then the grain was gently beaten out of the ear with a stick. This method ensured that neither the grain nor the straw was damaged.[5]

The most common method of threshing grain in Ireland from early times was with the flail, or 'súiste' (Pl. 4.1). Flail threshing apparently originated in Roman Gaul in the third or fourth century A.D. and spread throughout Europe and many parts of Asia. Evidence that the flail was used in Ireland at least before the eight century is provided by the Brehon Laws, which deal with compensation for injury arising from flail threshing.[6] The flail was still in common use in Ireland up to the beginning of the twentieth century.[7] This instrument was made from two straight sticks, tied together by a flexible joint. The sticks, known as the handstaff and the striker, were about one and a half inches and one inch in diameter, respectively, and were usually between three and four feet in length. The handstaff, also known as collop, 'colapán', and 'lamhchrann', was usually made from unpeeled hazel, pine or oak, while the striker, known variously as 'buailteán', 'buailcín', and souple, was made from holly, blackthorn, whitethorn or ash. The tying, or 'iall', thong, 'gad', hanging, tug, mid-kipple, hooden, middling, could be made from animal skin, including sheep skin and eel skin, flax cord, willow withies or straw rope.[8]

Pl. 4.1 Flail used in the mid Cork area (Courtesy of The Mills Agricultural Museum).

A variety of flail types existed in Ireland, distinguished by the different ways in which the sticks were tied together (Fig. 4.1). The main types were the 'double-loop' flail, predominant in Connaught, Clare and south-west Munster and found with other types in Ulster and east Munster; the 'eye' flail, common in Donegal and sometimes found in other parts of Ulster and in Connaught and Munster; the 'hole' flail and 'double-hole' flail found only in Ulster; and the 'cap' flail, with a distribution in Leinster and adjacent parts of Ulster and Connaught, as well as east Munster. Sub-types have also been identified, that of the double-loop flail found in County Limerick and surrounding areas and that of the cap flail found in Leinster and east Munster.[9]

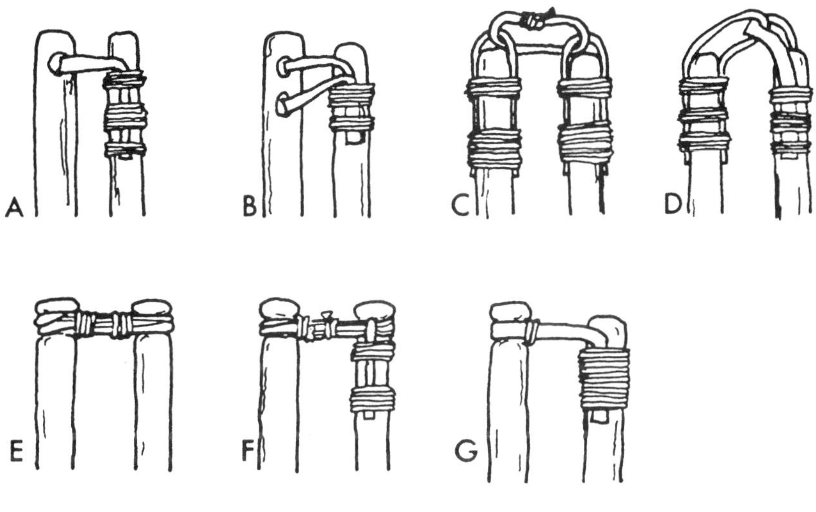

Fig. 4.1 Irish flail types: (A) hole flail - Ulster; (B) double-hole flail - Ulster; (C) cap-flail - Leinster, adjacent parts of Ulster and Connaught, and east Munster; (D) sub-type of cap-flail - Leinster and east Munster; (E) double-loop flail - Connaught, Munster, and Ulster; (F) sub-type of double-loop flail - Munster; (G) eye-flail - Donegal (After Bell, J. and Watson, M., 1986, p. 207).

Though a seemingly simple implement, much skill and energy was needed to thresh successfully with the flail. The sheaves were laid on the floor in front of the flailer with the ears pointing in towards one another. With the handstaff held in both hands, the whole flail was swung so that the striker was brought down forcefully on the sheaves, thus separating the grain from the ears.

Some flails were joined in such a way that the striker could swing freely around the handstaff, while others were not. In the latter case, the handstaff was turned in the hands as the flail was swung. Flail threshing could be carried out by one man, who could thresh up to four sheaves at once, or co-operatively by up to four men. Where there was more than one worker, good timing and rhythm were essential to prevent injury from the flail.[10]

Flail threshing was carried out in a variety of locations: in the open air, especially in the north, west and south-west of the country and on many small farms until recently; in the barn on larger farms particularly in Leinster; and sometimes in the farmhouse kitchen. In the early nineteenth century, the middle of the public road was often used. Whatever area was chosen, it was common to place a threshing board underneath the sheaves. This could be a door or a specially made smooth board, usually six feet square. Another type of threshing floor was one of prepared clay. A strong linen cloth was often spread over the threshing surface to facilitate removal of the grain and to keep it clean. A surface that would 'bend to meet the flail' was considered to be a good one. Noisy threshing was desired and the threshers would stamp their feet to the blows of the flail. A floor under which a horse's skull was buried was favoured since the sound of the threshing would be magnified by the skull.[11] One Irish proverb sums up the essentials of good threshing:

> "A hazel handstaff
> and a holly striker,
> a springy floor
> and a blow against the sheaf".[12]

Threshing Machines

The first attempts to develop machines for threshing grain were made in Scotland from the 1730s onwards. Machines which incorporated flails as well as those which rubbed the grain from the ear were invented. However, these early designs were unsuccessful and it was not until 1786 that the first practical threshing machine was developed by Andrew Meikle of East Lothian. His design laid the basis for subsequent developments. In this machine the grain was beaten rather than rubbed from the straw by the process of feeding sheaves between a rotating drum with four lengthwise bars and an outer metal casing. Water power was used to turn the drum inside the casing. Later machines included revolving rakes which raked off the straw.[13]

By the beginning of the nineteenth century, threshing machines had been installed on a number of large farms in Ireland, and some of these were powered by horses. Installation continued in the early decades of the century and included Scottish and Irish-made machines. However, these machines were confined to large farms since it was only those with substantial financial resources that could afford to acquire them and also to provide the motive power for them. Construction of either a water-wheel or a covered horse-engine was necessary. Horse-engine houses were a distinctive architectural feature on the landscape. Their circular or polygonal shape was dictated by the fact that the horses driving the engine walked in a circle, harnessed to a long arm which was attached to a central vertical mechanism. These houses, which were necessary to protect the mainly timber engines from the weather, were covered by high-pitched roofs of a conical or pyramidal shape. The early engines were heavy, with an overhead drive and a system of gearing which put considerable strain on the two to six horses required to power them. However, by the early 1840s, lighter and smaller horse engines made of cast metal had been invented (Fig. 4.2). The drive from this new type of engine to the thresher came out from beneath the gearing at the centre. The drive shaft was usually either enclosed in a ramp or in a stone-lined channel sunk in the ground underneath the horse walk. The fact that the new engines were constructed of metal rather than timber ruled out the necessity for building covered engine houses. This development of the smaller horse-engine made mechanized threshing more financially accessible to smaller farmers and was to have an important impact on its later diffusion to even very small farms in Ireland. Horse-driven machines significantly speeded up the job of threshing, being apparently five to six times quicker than the flail.[14]

Development of the horse-engine was paralleled by that of the barn thresher. Having begun production in 1837, Pierce's of Wexford quickly became the largest manufacturer of threshers in Ireland. In 1860, Kennedy's of Coleraine began making machines and became the leading manufacturers in the north of the country. By this time, machines were being made by many smaller foundries. Threshers began to be produced in a variety of sizes and designs (Pls. 4.2, 4.3

and 4.4). Two types of drums were available, the 'beater' drum which was based on Meikle's design, and a 'peg' drum which featured rows of short iron pegs.[15] The specific problem of threshing barley was also addressed. The removal of the long spines or 'awns' found on the heads of this crop required an operation that was usually separate from threshing. This process, known as hummelling, awning or avelling, was only successfully carried out by a hummeller, an implement commonly comprising a square metal plate of parallel iron strips to which was fitted a shaft with a T-shaped handle. By the 1830s, modification were being made to threshers so that hummelling could be done as part of the overall threshing operation.[16]

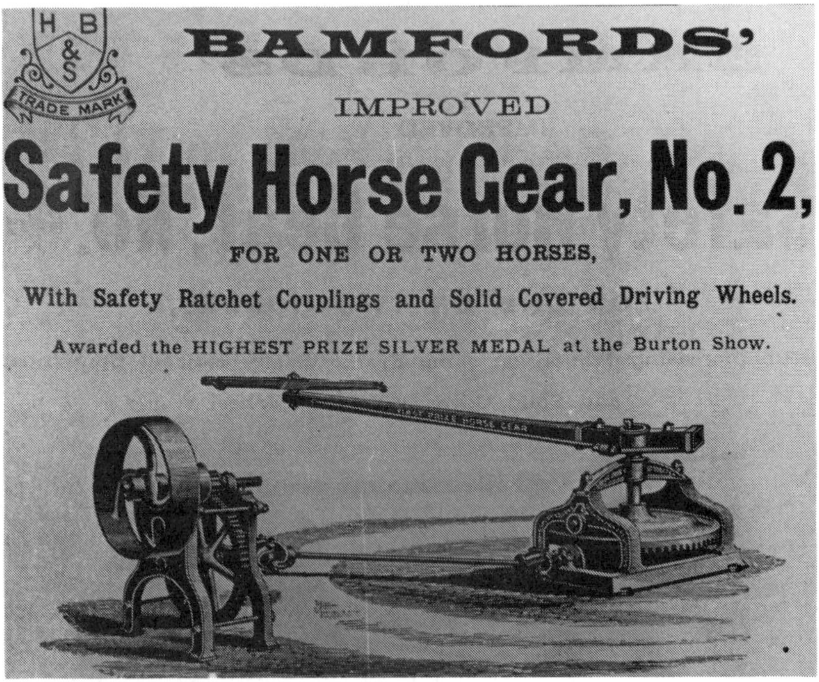

Fig. 4.2 Bamfords' Horse Gear (*Bamfords' General Catalogue of Agricultural Machines*, No. 27, Nov., 1894).

Pl. 4.2 Drum thresher used in north Cork in the later decades of the nineteenth century. This type was popular throughout the country. (Courtesy of Robert Troy).

Pl. 4.3 A rare thresher type used in north Cork, 1860s-1880s. Unlike the more common drum thresher, it featured 'walkers' which propelled the straw out of the machine. (Courtesy of Robert Troy).

Pl. 4.4 Reed thresher used in north Cork, 1860s-1880s. This type was used for making reed for thatching. (Courtesy of Robert Troy).

In the mid nineteenth century, two developments brought mechanized threshing within reach of even the smallest farmer. Very small machines which were turned by hand came to the attention of the Irish farmer and these, made by Barrets of Reading in Yorkshire, became very popular during the latter half of the century. Machines which could be operated by pedals as well as by hand were also developed and came into use (Pl. 4.5). The second development was the provision of threshing systems in portable form (Fig. 4.3). This was made possible by the earlier introduction of the lighter and small horse-engines. Garrett's horse-operated portable threshing machine was exhibited in London at the Great Exhibition of 1851 and two years later similar threshers were exhibited at the Irish Industrial Exhibition in Dublin. These machines could be hired out and moved from one farm to another, and they became a significant feature of Irish threshing in the second half of the nineteenth century.[17]

Pl. 4.5 Pedal and hand thresher used in mid Cork (Courtesy of The Mills Agricultural Museum).

Fig. 4.3 Horse-powered portable threshing machine manufactured by Barrett, Exhall and Andrewes of Reading (*The International Exhibition of 1862*, p. 5).

The use of portable machines became much more widespread, however, with the introduction of steam power to threshing. Portable steam engines, pulled by horses, came into use as the power source for greatly enlarged threshing machines (Figs. 4.4 and 4.5). A more significant development, however, was the invention of steam traction engines (Fig. 4.6). These engines could pull as well as power threshers. The Lincolnshire firm of Tuxfords of Buxton have been attributed with the development of steam threshing in 1842. From the late 1850s, the demand for steam threshers grew and by the end of the century there were over five hundred of these machines or 'mills' in Ireland. The 'Threshing Set' became a very important feature of the threshing scene, and eventually included the engine, thresher, straw elevator, baler and trusser. The most common machines used in Ireland were made by Ransome, Marshall, Garvey, Boyd, Clayton & Shuttleworth, Ruston, Creighton, Fisher Humphries, and Foster. Since few farmers could afford to invest in a steam thresher, a substantial business of hiring out 'threshing sets' developed in Ireland. These 'sets' travelled from farm to farm during the months of August, September and October (Pl. 4.6). Large fleets of 'sets' were operated by some contractors, particularly in counties Carlow, Kilkenny, Kildare, Cork and Wexford. One of the largest operations, run by Captain Henry Osborne from Drogheda, was made up of twelve complete sets and seventeen engines. Smaller operators were common throughout the country. Along with paying for the hire of the 'set', the farmer was commonly obliged to provide board and lodging for the operator until the whole crop was threshed. The farmer supplied the coal for the engine and was expected to refill it before it moved on to the next farm .[18]

Steam threshing continued in Ireland until it was replaced by tractor power in the 1940s (Pl. 4.7). In 1939, there were just over two thousand tractors in the twenty-six counties, but the number grew dramatically after World War II to 13,569 in 1950 and to 43,697 in 1960. In the harvesting and threshing of grain, tractor power was in turn replaced by the combine harvester in the 1960s, the numbers of which had increased from just 366 in 1939 to 4301 in 1960 (Pl. 4.8).[19]

While there were some reservations expressed during the nineteenth century about the efficiency of threshing machines, their major contribution was the speed with which they could thresh. Work which was often carried out throughout the winter months by traditional methods, especially the flail, could be done in a few days by a threshing machine.[20] Yet, the flail was not readily abandoned in Ireland. This was mainly due to the abundance and low cost of farm labour in pre-Famine times, compared to the high cost and bulkiness of the earlier threshing machines. However, a reduction in the number of farm labourers in the second half of the nineteenth century together with the developments which made mechanized threshing more generally available meant that the traditional methods of threshing went into decline.[21] The flail continued to be used occasionally for specific purposes, for example, to obtain top quality straw, small quantities of seed for the crop of the following year, or fresh feeding straw.[22]

Fig. 4.4 Portable steam engine (*Ransomes, Sims & Jefferies' Illustrated Catalogue of Steam Thrashing Machinery*, July, 1921).

Fig. 4.5 Ransomes' Threshing Machine (*Ransomes, Sims & Jefferies' Illustrated Catalogue of Steam Thrashing Machinery*, July, 1921).

Fig. 4.6 Steam traction engine (*Ransomes, Sims & Jefferies' Illustrated Catalogue of Steam Thrashing Machinery*, July, 1921).

Pl. 4.6 A threshing set on its way from one farm to the next in mid Cork, 1930s (Photograph courtesy of Richie Walsh/Seán McSweeney).

Pl. 4.7 Threshing with tractor power in mid Cork in the 1940s.

Pl. 4.8 The first combine harvester to be used in Ireland, 1934 (Upton Steam Rally, 1995).

Threshing and Co-operative Labour

With the widespread adoption of threshing machines, threshing became one of the events of the farming year when co-operative effort was most required. Though flail threshing did entail co-operation in some cases, it was not necessary. However, mechanized threshing, particularly steam threshing, necessitated a large work force so that most could be made of the speed at which the machine could thresh (Pl. 4.9).[23] Along with the driver of the thresher who tended the engine, up to twenty people were involved in the threshing operation. Typically, two men on top of the stack - which had been built at the harvesting stage - pitched sheaves on to the platform of the drum of the thresher. Another two men or boys cut the tying bands from these sheaves and then passed them to the feeder. On equal footing with the driver, the feeder usually determined the pace of threshing. He was responsible for feeding the sheaves into the exposed revolving drum. This was an acquired skill, and was a dangerous job demanding full concentration. Up to three men bagged the grain and carried it to the barn. Sacks came in different sizes depending on the crop - oats, barley and wheat sacks weighing fourteen, sixteen and twenty stone, respectively. In the

Pl. 4.9 Threshing day in mid Cork, *c.* 1910. The 'car' or cart close to the steam engine was used for carrying oil lubrication for the thesher, to which it was attached when being transported (Photograph courtesy of Jimmy Moynihan).

era before balers and trussers, threshed straw was built into a stack or rick. Emerging from the back of the drum, this straw was pitched by three men to those building the rick (Pl. 4.10). This latter task, and the subsequent job of securing the rick with straw or 'súgán' ropes, required the labour of about six men - usually skilled rick-makers from the local area. One worker removed the chaff and this was considered to be the dirtiest job of the threshing.

The threshing day was an important social occasion, and often more people than required were present. That the co-operative work of the day went hand-in-hand with considerable supplies of alcoholic drink is attested to by many oral accounts, including one from County Cork which states that:

> "the digle a stroke they'd do
> if they hadn't porter"[24]

The highlight of the day was the communal meal which was prepared by the women for this 'meitheal' or team of workers, and which was often followed by festivities, sometimes known as 'the time' or 'the social'.[25] The co-operative and social nature of the threshing day, and indeed of the preceding harvesting operation, was effectively brought to an end with the introduction of the combine harvester which cut, threshed and winnowed the grain in one swift operation.

Pl. 4.10 Workers piking straw as it emerges from the thresher, in mid-Cork, *c.* 1950.

GRAIN WINNOWING

Grain seed was suitable for many purposes only if the husks or shells were separated from the threshed grain. Known as winnowing, this process was traditionally carried out by hand - and in some areas into the twentieth century (Pl. 4.11). It simply involved shaking the grain from a riddle or tray in the open air on

Pl. 4.11 Winnowing grain by hand, Aran Islands, County Galway, *c.* 1930 (Photograph courtesy of the National Museum of Ireland).

a windy day. The heavier grain seed fell straight down on to a sheet spread in front of the winnower while the lighter husks were blown away. Winnowing was carried out on a road or in a field. High exposed places, sometimes known as 'Shilling' (shelling) Hills, were favoured. Riddles were made from strips of wood, usually ash, which were woven and attached to a circular frame made of the same material. The size of the meshes depended on the type of grain being winnowed. Winnowing trays were also used. Those known as 'blind sieves' were similar to riddles except that the strips were woven very tightly to prevent holes. Trays were also made from sheep or goat skin stretched over a thin, wooden, circular rim. This type of tray was known as a 'dallán', a name also given to the 'blind sieves', and in a smaller version was known as a wight or wecht in the north and as a borrane or 'bodhrán' in the south. The latter is a name also given to a popular percussion instrument of a similar make. Winnowing with skin trays was common throughout Ireland. Once winnowed, the grain was bagged and the chaff which was of little or no value was discarded.[26]

A machine for winnowing grain was developed by James Meikle, the father of Andrew Miekle, early in the eighteenth century. Grain fed into this machine, or 'barn fan', was separated from the chaff by falling through an artificial current of air which was created by turning canvas sails. A series of sieves also cleaned the seed. A more compact version of Meikle's machine was built in 1770 by the Englishman James Sharp. This had wooden fans which were turned by a handle at the side of the machine. Barn fans were further improved so that by the middle of the nineteenth century grain of different weight and quality was divided and delivered through different spouts. Machines were fitted with riddles whose sizes varied according to the type of grain being winnowed. Barn fans were relatively cheap to purchase, easy to use because of their gearing system, and produced faster and better results than hand winnowing. Consequently, they were adopted in some parts of Ireland by the beginning of the nineteenth century and usually preceded that of mechanized threshing by a generation or more (Pl. 4.12).[27]

Threshing and winnowing remained separate operations into the nineteenth century. Hand-threshers and most barn-threshers did not winnow grain. However, by the 1830s and particularly with the introduction of steam power, machines which combined both processes were being developed. In the following decades, a number of refinements were made, including the mechanical bagging of grain. However, on some farms, particularly on those which did not avail of steam threshing, winnowing continued to be done by hand or by use of the barn fan.[28]

Pl. 4.12 A restored winnowing machine, manufactured by Josiah Le Butt, Bury St. Edmunds (Courtesy of Robert Troy).

STORAGE OF GRAIN

On some farms in Ireland, grain was stored unthreshed in stacks, especially before the introduction of mechanical threshing. The threshing was carried out by flail over the winter and spring months and was a valuable source of employment for rural labourers.[29] On other farms, grain was threshed shortly after harvesting. Threshed grain that was not destined for market was stored on the farm in a number of ways. In the nineteenth century, grain was usually stored in a barn on larger farms and by the middle of the present century barns had become a common feature on many farms throughout Ireland. Up to that time, however, many smaller farms did not possess such a building, and other storage methods were employed. One such method, known in Ireland for several hundred years, was the use of straw or 'súgán' ropes to build an outdoor cylindrical granary (Fig. 4.7). A layer of furze, twigs or brambles, or a platform of stone slabs, was covered with straw. The latter formed the base of the granary, which could have a diameter of between six and twenty-six feet. Thick coils of rope, often made from the newly threshed straw, were wound round in a circular fashion to form the walls. As it was being built, the granary was filled with grain, which was left unwinnowed. Once the sides had reached a height of five or six feet, a water-tight roof of thatch was added, bringing the total height to between seven and twelve feet. Up to four people were involved in the construction of this straw rope granary. Known also as 'foirin', 'siogog', 'doimhineog', 'sop', 'darnaol' and 'lochtan', this type of granary was a temporary structure. It was opened and dismantled in the spring when the winnowing of grain it had held was carried out. Straw rope granaries were recently recorded in counties Cork and Kerry, but were probably known in many other parts of Ireland in earlier times.[30]

Until the nineteenth century, the flail was the primary tool used to thresh grain. Though a simple implement in its variety of forms, its successful use demanded much skill and energy. Mechanized threshing came into being in the early nineteenth century, as did mechanized winnowing. It was the introduction of steam power that speeded up the diffusion of mechanized threshing. This brought about major change, with threshing of the crop being carried out in a few days rather than over a few months, as was traditionally the case. With the introduction of the 'Threshing Set' came the greatly increased need for co-operative help, which in turn heralded the beginning of a major social occasion in the farming calender.

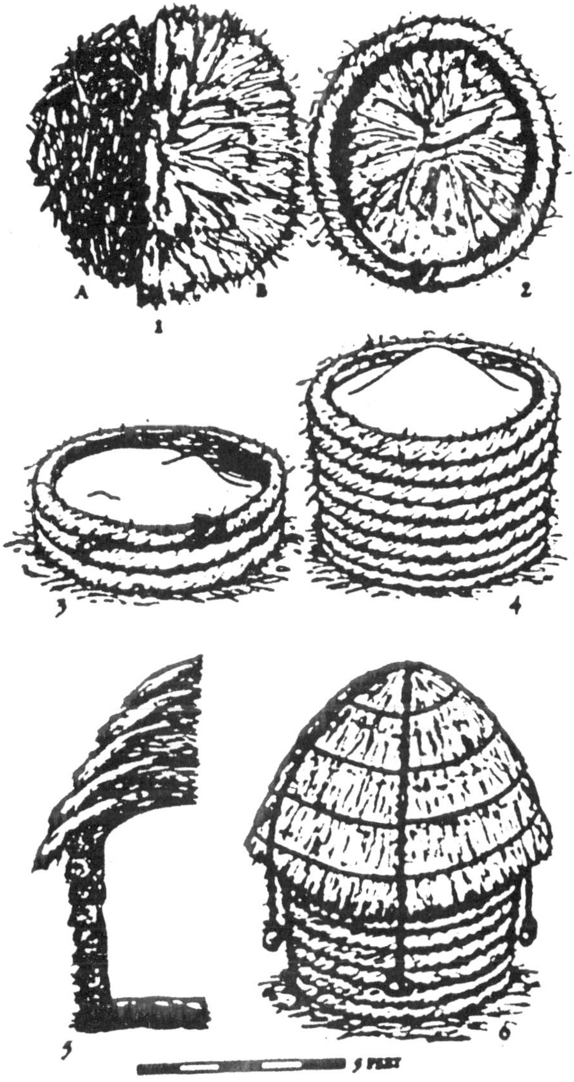

Fig. 4.7 Straw-rope granary (After Sigaut, F., 1988, p. 14).

References

1 Pollack, V. (1988) 'Work on the Land', 20-1; O'Neill, T. P. (1977) *Life and Tradition in Rural Ireland,* 39; Collins, E. J. T. (1972) 'The Diffusion of the Threshing Machine in Britain, 1790-1880', 20-1; Bell, J. and Watson, M. (1986) *Irish Farming - Implements and Techniques 1750-1900,* 217.

2 Bell and Watson, op. cit., 220-4 and 224; Blandford, P. W. (1976) *Old Farm Tools and Machinery,* 130.

3 Ó Danachair, C. (1955) 'The Flail and other Threshing Methods', 7-8; Danaher, K. (1972) *The Year in Ireland,* 167; Bell and Watson, op. cit., 205.

4 Ó Danachair, C. (1955) op. cit., 8-9; Bell and Watson, op. cit.

5 Ó Danachair, C. (1955) op. cit., 9; Danaher, K. (1962) *In Ireland Long Ago,* 93.

6 Ó Danachair, C. (1955) op. cit., 13; Danaher, K. (1962) op. cit., 96; Gailey, A. (1984) 'Introduction and Spread of the Horse-powered Threshing Machine to Ulster's Farms in the Nineteenth Century: some aspects', 37-8.

7 Ó Danachair, C. (1970b) 'The Flail in Ireland', 55.

8 Ó Danachair, C. (1955) op. cit., 9-13; Evans, E. E. (1957) *Irish Folk Ways,* 213.

9 Ó Danachair, C. (1955) op. cit., 11-12; Ó Danachair, C. (1970b) op. cit., 50-5; Bell and Watson, op. cit., 206-7.

10 Ó Danachair, C. (1955) op. cit., 9; Bell and Watson, op. cit., 208; Gailey, A. op. cit.

11 Ó Danachair, C. (1955) op. cit., 9-10; Bell and Watson, op. cit., 208; Evans, E. E. op. cit., 215-6; Pollack, V. op. cit., 19-20.

12 Danaher, K. (1962) op. cit., 93-4

13 Bell and Watson, op. cit., 209; Gahan, W. (1956) 'The Origins of Crop Husbandry', 254; Blandford, P. W. op. cit., 128; Gailey, A. op. cit., 39

14 Bell and Watson, op. cit., 209-14; Gailey, A. op. cit., 39 and 45.

15 Bell and Watson, op. cit., 212-3; Gahan, W. (1956) op. cit.

16 Bell and Watson, op. cit., 218.

17 Ibid., 214-7; Gailey, A. op. cit., 45; Blandford, P. W. op. cit., 129.

18 Bell and Watson, op. cit., 217; Pollack, V., op. cit.,21;Neill-Watson, J. (1983) *A History of Farm Mechanisation in Ireland 1890-1990,* 1-5; O'Mahony, S. (1993) *A Gate to the Past,* 178-180.

19 Gillmor, D. A. (1977) *Agriculture in the Republic of Ireland,* 73; Neill-Watson, J. op. cit., 2-3; Crowley, F. (1979) *In West Cork Long Ago,* 80-2; O'Neill, T. P. op. cit., 88.

20 Bell and Watson, op. cit., 218-9; Collins, E. J. T. op. cit.;

21 Gailey, A. (1990) '"...such as pass by us daily...": The Study of Folklife', 12-3; Bell and Watson, op. cit., 219-220;

22 Pollack, V. op. cit., 20; Collins, E. J. T. op. cit., 22.

23 O'Dowd, A. (1981) *Meitheal,* 75 and 101; Bell and Watson, op. cit., 217

24 O'Dowd, A. op. cit., 112.

25 Ibid.; Danaher, K. (1962)op. cit.; O'Mahony, S. op. cit., 176.

References

26 Bell and Watson, op. cit., 220-2; Evan, E. E. (1942) *Irish Heritage*, 124; Evan, E. E. (1957) op. cit., 213; Crowley, F. op. cit., 86-7; O'Neill, T. op. cit., 39; Pollack, V. op. cit.

27 Bell and Watson, op. cit., 222-3; Blandford, P. W. op. cit.; Gailey, A. (1984) op. cit., 39.

28 Bell and Watson, op. cit., 224; Blandford, P. W. op. cit., 130.

29 Gailey, A. (1984) op. cit., 38.

30 Bell and Watson, op. cit., 224-5; Crowley, F. op. cit., 80-4; Siguat, F. (1988) 'A Method for Identifying Grain Storage Techniques and its Application for European Agricultural History', 14; O'Dowd, A. op. cit., 76.

CHAPTER 5

GRAIN MILLING FROM EARLIEST TIMES TO THE SIXTEENTH CENTURY

An essential stage in the transformation of the grain crop into food suitable for human and animal consumption was the grinding and milling of the grain seed. This was possible once threshing of the crop had been completed. Querns were used from early times to carry out this task on a small scale. Large-scale production was made possible with the introduction of water-powered grain mills from early medieval times. Milling, even when mechanized, was an involved process, and more so in Ireland than elsewhere where the damp climate necessitated the drying and hardening of the grain seed before the process could commence.

HAND-QUERNS

With the introduction of grain cultivation in Ireland before 3000 B.C. by the Neolithic settlers, methods of grinding the hard grain seed into food were developed. The earliest implement to be used was the saddle quern (Fig. 5.1). This consisted of a rough stone, about twenty-four inches long, with a hollow in the centre, and a smaller hand-held rubbing stone of about ten inches in length. The latter was used to pound the grain which was placed in the hollow of the larger stone.[1]

By the early Iron Age, the rotary quern had been developed and in comparison to the saddle quern was capable of producing a finer product in a shorter amount of time (Pl. 5.1). This quern consisted of two flat circular stones, each of about twenty inches in length. One stone was placed on top of the other and both were joined by a central wooden spindle. The lower stone remained in a stationary position. The upper stone was turned by means of a handle attached near its edge. As this stone was being rotated, grain was fed by hand through a hole at its centre. In many examples, the grain was distributed by means of channels cut into the lower side of the upper stone (Pl. 5.2). The ground product emerged from between the edges of the two stones.[2]

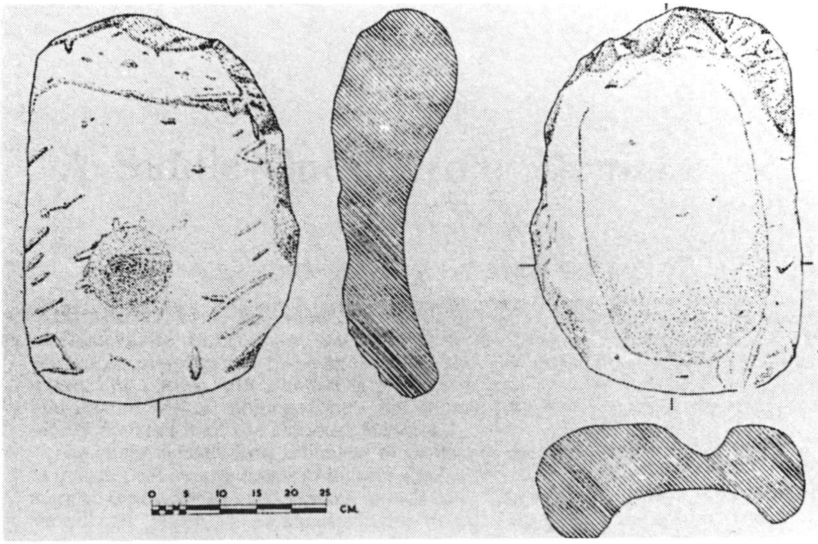

Fig. 5.1 Saddle-quern discovered at Donickmore, County Cork (After Barber, J. and Deady, J., 1970, p. 55).

Pl. 5.1 Rotary quern common to the south and west of Ireland (Photograph courtesy of Donal J. O'Sullivan).

Pl. 5.2 Rotary quern showing channels on lower side of upper stone (Courtesy of Lee Valley Open Farm).

Though water-powered grain mills were introduced into Ireland by the early medieval period, archaeological and written evidence point to the continued use of the rotary quern for small-scale production, even into the present century. The sustained use of the rotary quern in early Christian and early medieval Ireland is indicated by frequent finds on sites dating from these periods, as well as by numerous references to this type of quern in the Brehon law tracts and other literature. In early Irish society, the onerous task of grinding grain with a quern was carried out by women, specifically by the 'cumal' or bondmaiden.[3] It is recorded in the twelfth century text, *Cogadh Gaedhel re Gallaibh,* that Brian Boru made quern-maids of his Viking female prisoners.[4]

Several examples of another type of quern have been found in the northern half of Ireland. Known as the beehive quern, this type was roughly circular in shape with a diameter of about twelve inches. It consisted of a flat lower stone and a somewhat smaller beehive-shaped upper stone. The latter stone had a funnel-shaped central perforation through which the grain was poured as the stone was being rotated by means of a wooden handle. Spiral decorations on some examples, along with their northern distribution, suggest to archaeologists that these querns were introduced into Ireland in the Iron Age by the La Tene Celts, between the second century B.C. and the fourth century A.D.[5]

WATER-POWERED GRAIN MILLING

The introduction of water-powered grain mills into Ireland can be dated to at least the early seventh century A.D. That these mills played a very significant role in the subsistence agricultural economy of early and medieval Ireland is clearly indicated by the several sites discovered throughout Ireland by archaeologists, and by the evidence of an important site in Little Island, County Cork, which shows that millwrighting was an advanced stage in seventh-century Ireland. The construction of a water-mill required considerable investment, and the existence of these mills in considerable numbers points to relatively large-scale production.[6] The Brehon law tracts confirm the archaeological evidence. *De ceithri slichtaib Athgabala*, a legal tract dating from the seventh century or earlier, describes several parts of the mill. The *Cain Domnaig* or Law of Sunday, dating from the eight and ninth centuries, forbids the grinding of grain in a mill or quern on Sunday. Probably the most important tract is the eight-century *Coibnes Uisci Thairidne*. This deals with water rights and watercourses and is the most complete collection of milling law in Europe from this period.[7] It also provides an insight into mill ownership, indicating that the sharing of mills was common in early medieval Ireland, particularly among less well-off members of the community. The communal aspect of the grain mill was not confined to the early medieval period. During the Anglo-Norman period, the manor mill, which was the property of the lord, was an important element of the manorial centre. The tenants of the manor were obliged, under the Norman manorial system, to grind their grain at the mill and also to provide labour for its working and maintenance. In the pre-Reformation period, monastic sites often included a mill which served the local community as well as the monastery itself, with some of these mills even operating to the level of the manor mill.[8]

Both horizontal and vertical-wheeled water-mills were in use from the seventh century, though the former was much more common until well into the sixteenth century. The horizontal-wheeled water-mill was a significant technological development in the history of Irish milling. It reduced considerably the burden of grinding grain, which in the early medieval period included wheat, barley, oats and rye. The output of a small horizontal-wheeled mill has been estimated to be in the region of forty to fifty pounds of meal per hour compared to that of a hand-quern operated by two slaves of sixteen pounds. Mechanisms for adjusting the stones of the water-mill meant that the flour or meal produced was of a much finer quality than that ground by a hand quern, which had no such facility.[9]

The horizontal-wheeled mill consisted of a two-storey wooden wheelhouse. The lower section contained the timber horizontal wheel while the upper floor held the millstones (Pl. 5.3a). The water supply was usually provided by a nearby fast-flowing stream. In most cases, a small reservoir or mill-pond was created by the construction of a low bank of earth, stones or clay over this stream, from which a feeder channel or mill-race was dug in order to direct the water to the mill-house. Sluice gates were used to cut off this water supply when the mill was not in operation (Pl. 5.3b). The building of a mill-race required considerable resources and was probably a common activity in early medieval society. Water from the mill-race was directed into a hollowed-out wooden flume or chute. The latter was splayed internally, ensuring that a water jet of sufficient pressure was directed onto the dished vanes or paddles of the wheel. The water-wheel, which was mounted on a cup-shaped pivotstone, was forced to rotate. The timber drive which connected the wheel to the upper millstone or 'runnerstone' on the floor above was direct, that is, gearing was not involved, and so every turn of the wheel resulted in a corresponding turn of the millstone. An important discovery of a complete mill-wheel with paddles was made in a recent excavation of a mill site at Cloontycarthy, near Macroom, County Cork, which dates to about 833 A.D. (Fig. 5.2).[10]

(a)

(b)

Pl. 5.3 Reconstruction of a horizontal-wheeled water-mill which was located at Mashanaglass, near Macroom, Country Cork: (a) upper and lower storeys of the wheelhouse; (b) feeder channel and sluice gates (Bunratty Folk Park, County Clare).

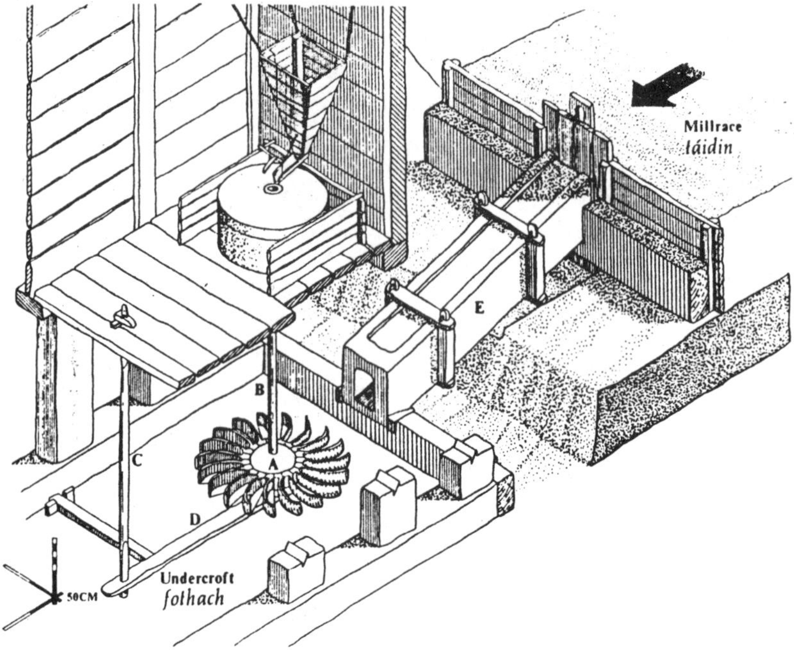

Fig. 5.2 Reconstruction of a horizontal-wheeled water-mill at Cloontycarthy, County Cork, *c.* 833 A.D. (After Rynne, C., 1990, p. 22).

The runnerstone in these horizontal-wheeled mills had a diameter of about twenty-four to forty inches, while the lower stationary stone or 'bedstone' was about twenty-two to thirty-nine inches in diameter. Sandstone was the most common material used in the construction of these millstones. The first step in the milling process was the feeding of the grain into a perforation at the centre of the runnerstone. Archaeological and documentary evidence indicate that this was done automatically by means of a hopper (see Fig. 5.2). The grain was then ground between the two stones. The lower side of the runnerstone and upper side of the bedstone were dressed, that is, they had grooves chiselled into them. This was to ensure that the grain was properly directed toward the stones' edges, from which the milled product emerged. Milling involved two stages: the husks were separated from the grains and then the kernels were ground into flour or meal. In water-mills, these stages were facilitated by the use of a tentering mechanism which allowed for precise adjustments of the millstones.

Several horizontal-wheeled mill sites have been discovered in Ireland. Due to the preservation of ancient timbers, more is known about early medieval Irish mills than about those of the remainder of Europe. For example, discoveries of mill sites of this period in County Cork alone exceed that of western continental Europe. Nineteen Irish sites have been accurately dated by dendrochronology to the period 630 to 926±9 A.D. Five of these are located in County Cork, while at least an additional nineteen sites in this county are likely to date also from the early medieval period. One of the County Cork sites, Little Island in Cork Harbour, is of major importance for a number of reasons. Not only is it the earliest site in Ireland to be accurately dated by dendrochronology, but it is also the earliest in Europe. The mill seems to have used two wheels, making it at present the earliest example of its type in either Europe or Asia (Fig. 5.3). As well, this mill, along with a vertical-wheeled mill which was also discovered at the site, seems to have been powered not in the most usual way by a stream but rather by tide power. The incoming tide would have filled the millponds, the mills would have been worked when the tide ebbed, and the ponds would have been filled again by the flow tide. This discovery invalidates the previously-held belief that tide mills had their origins in the high Middle Ages, and clearly establishes a much earlier date.[11]

A commonly-held assumption has been that the vertical-wheeled water-mill was introduced into Ireland by the Cistercians and the Anglo-Normans during the twelfth century. Recent archaeological developments have indicated otherwise. The vertical-wheeled mill excavated at Little Island, County Cork, has been dated by dendrochronoly to 630 A.D. - the same date as the nearby horizontal-wheeled mill found at this site - and so far is the earliest example of its type in Ireland (Fig. 5.4). At Morrett in County Laois, a mill which was thought at the time of its excavation in the 1950s to be a horizontal-wheeled mill is now known to be a vertical-wheeled water-mill. This mill has been accurately dated to 770 A.D. As in the case of Little Island, an undershot vertical wheel was used in Morrett, that is, incoming water was directed at the lower part of the wheel. Thus, both horizontal and vertical-wheeled mills in Ireland date from at least the early medieval period, while the close association of both types at Little Island is the earliest known anywhere in Europe.[12]

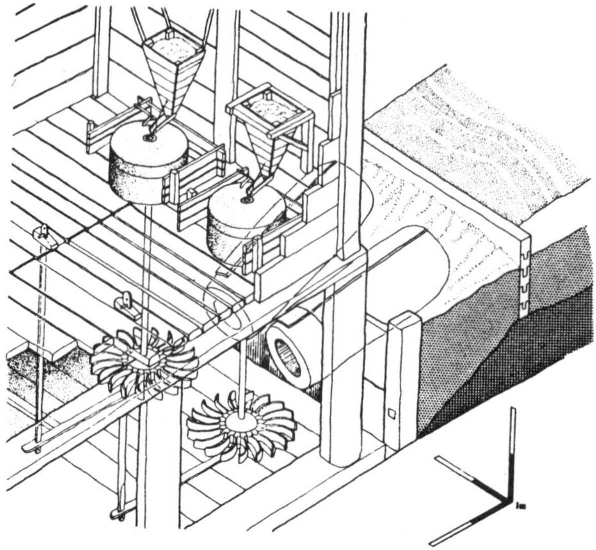

Fig. 5.3 Reconstruction of a seventh-century horizontal-wheeled water-mill at Little Island, County Cork (After Rynne, C., 1993, p. 26).

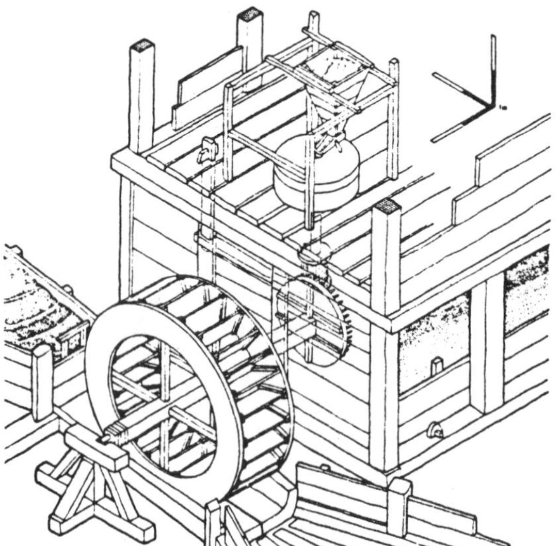

Fig. 5.4 Reconstruction of a seventh-century vertical-wheeled water-mill at Little Island, County Cork (After Rynne, C., 1993, p. 27).

The vertical-wheeled mill was usually powered, as in the case of most horizontal-wheeled mills, by water which was diverted from a nearby stream. However, the vertical position of the wheel in the former mill type meant that the driveshaft was set horizontally and thus power transmission to the millstones was indirect. In other words, right-angled gearing was required. Early mills powered by vertical wheels had an advantage over horizontal-wheeled mills in that it was possible to adjust the millstone speeds according to the power of the stream.[13]

There were a variety of vertical wheel types (Fig. 5.5). The oldest type, the undershot waterwheel, consisted of flat, open paddles, which dipped into the water flowing underneath, thus causing the wheel to turn. A later and more efficient development were the breast wheels. These were closed and consisted of buckets, the weight of the water in the buckets causing the wheel to turn. The water was made to impinge on the wheel's circumference - the term for each of the wheel types in this category being determined by the point of impact: low breastshot, just below the axle level, and breastshot, high breastshot and pitchback, all above the axle level. The final type was the overshot wheel. Water was discharged over the top of the wheel from a point past its apex. This caused the wheel to turn in an outward direction from the water delivery point, and differed from the other types in this respect.[14]

Advances in mill construction techniques during the Middle Ages meant that vertical-wheeled mills eventually replaced horizontal-wheeled mills in Ireland and most of northern Europe. However, this did not occur in Ireland until the early seventeenth century.[15] Thus, the horizontal-wheeled mill remained the dominant type in Irish society for at least one thousand years and was still in use in some remote parts of Ireland, especially in the north-east, even into the present century.[16]

DRYING OF GRAIN

The relatively damp climate of Ireland usually necessitated the drying and hardening of grain before milling. Experiments with rotary querns have shown that the milling of undried grain is difficult and ineffective, and consumes much time and labour. In early times, it is likely that drying was achieved by methods such as 'burning in the straw'. Such methods were suitable for small-scale grain processing. By the early historic period, grain-drying kilns had come into use. The many references to such kilns (áith) in the early literature point to their importance in the Irish landscape, as well as to the significance of grain growing in the economy. These kilns, according to the early Irish law tracts, were closely associated with water-mills, as well as flail threshing, and hence the processing of cereals on a larger scale.[17]

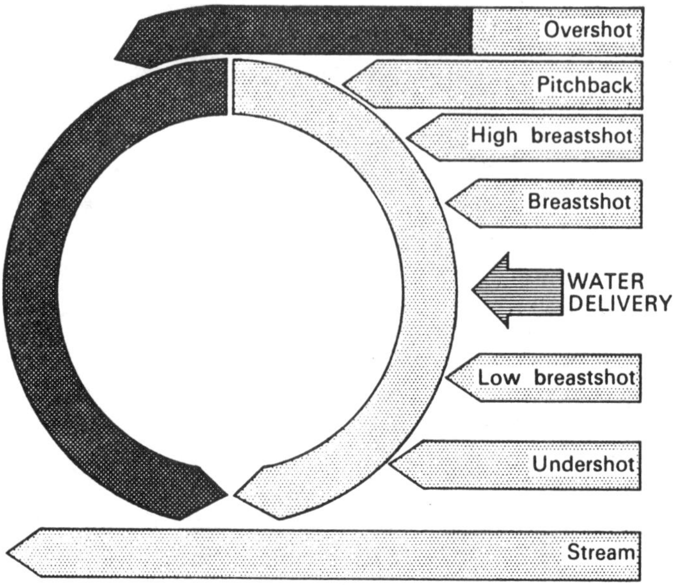

Fig. 5.5 Vertical mill-wheel types, distinguished by point of water delivery or contact (After McCutcheon, W. A., 1980, p. 233).

A variety of kiln types were known in Ireland. One early type consisted of a platform structure attached to a barn in which the other stages of crop processing were carried out. By early medieval times, free-standing kilns had come into use. One such type, known as the keyhole kiln, became the most common type to be used in Ireland, especially to the west and north-west (Fig. 5.6). It consisted of a circular drying chamber, into which a long flue led. This chamber was constructed in a rectangular-shaped stone-built structure which held and radiated the heat. It was covered with a timber thatch superstructure, which protected the grain. These kilns were multi-functioned. Along with drying or ripening the grain before milling and sometimes before threshing, they were also used to fumigate pests such as the grain weevil, to roast sprouted grain in the production of malt, and to reduce the moisture content of grain before storage.[18]

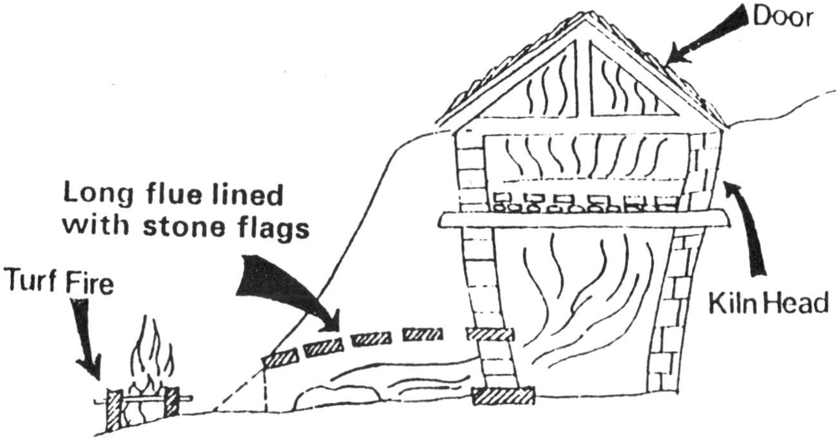

Fig. 5.6 Grain-drying kiln used in the Ballyhaunis area of County Mayo up to the 1840s (After Bowie, G., 1979, p. 8).

Like other aspects of cereal production, kiln-drying of grain usually involved a co-operative effort. The law tracts indicate that in the early medieval period grain-drying kilns and barns, along with mills, were shared by the less well-off farmers - though outright ownership of such structures was common among the more prosperous farmers. Free-standing kilns remained in common use up to the late eighteenth century and in some regions of the west of Ireland until the 1840s, with shared ownership continuing to be a feature.[19]

The milling of grain by hand-quern was burdensome. While rotary querns were in use up to the present century, it is little wonder that the milling process was mechanized relatively early. Water-powered grain milling was a vital aspect of the subsistence economy. While both horizontal and vertical-wheeled water mills were in use in Ireland from the early seventh century, it was the former which prevailed in the following millennium. Like mechanized milling, grain-drying kilns were an early feature of the economy, their introduction spurred on by the necessity for drying grain in Ireland's damp climate.

References

1 Buchanan, R. A. (1982) *Industrial Archaeology in Britian*, 193; Barber, J. and Deady, J. (1970) 'A Saddle-Quern from Donickmore, Co. Cork', 54-5; Flanagan, L. *A Dictionary of Irish Archaeology*, 179; O'Sullivan, D. J. (1995) *The Storyof Corn and Flour Milling In Ireland*, 2.

2 Evans, E. E. (1981) *The Personality of Ireland*,150; O'Sullivan, D. J. op. cit., 3; Flanagan, L. op. cit.

3 Rynne, C. (1990) 'Some Observations on the Production of Flour and Meal in the Early Historic Period', 20; Edwards, N. (1990) *The Archaeology of Early Medieval Ireland*, 63.

4 Ó Corráin, D. (1972) *Ireland before the Normans*, 46.

5 Flanagan, L. op. cit., 189; Raftery, B. (1984) *La Tene in Ireland*, 244.

6 Rynne, C. (1993) *The Archaeology of Cork City and Harbour from the Earliest Times to Industrialisation*, 25-30; McCutcheon, W. A. (1980) *The Industrial Archaeology of Northern Ireland*, 232.

7 MacEoin, G. (1981) 'The Early Irish Vocabulary of Mills and Milling', 13-4; Hamlin, A. (1981) 'Using Mills on Sunday', 11; Rynne, C. (1989a) 'Archaeology and The Early Irish Watermill', 110-1.

8 Edwards, N. op. cit.; O'Sullivan, D.J. op. cit., 86-7.

9 Rynne, C. (1989a) op. cit.; Edwards, N. op. cit.; Power, D. et al (1992) *Archaeological Inventory of County Cork*, 289; McCutcheon, W. A. op. cit., 274; Rynne, C. (1990) op. cit., 21.

10 Edwards, N. op. cit., 63-4; Rynne, C. (1989a) op. cit., 110-2; Rynne, C. (1989b) 'The Introduction of the Vertical Waterwheel into Ireland', 30; Rynne, C. (1990) op. cit.; McCutcheon, W. A. op. cit., 232.

11 Holt, R. (1990) 'Milling Technology in the Middle Ages', 52 and 56; Rynne, C. (1992) 'Milling the 7th century - Europe's Earliest Tide Mills', 24 and 54; Rynne, C. (1990) op. cit.; Edwards, N. op. cit.; Rynne, C. (1989a) op. cit., 111; Rynne, C. (1993) op. cit., 29.

12 Rynne, C. (1989b) op. cit., 26 and 30; Rynne, C. (1989a) op. cit., 113-4; Rynne, C. (1992) op. cit., 24; Rynne, C. (1993) op. cit., 28.

13 Rynne, C. (1989a) op. cit., 113.

14 McCutcheon, W. A. op. cit., 233; O'Sullivan, D. J. op. cit., 10.

15 Buchanan, R. A. op. cit., 236; Rynne, C. (1989a) op. cit., 113; McCutcheon, W. A. op. cit., 232

16 Edwards, N. op. cit., 63; Rynne, C. (1993) op. cit., 25; Bowie, G. (1979) 'Corn Drying Kilns, Meal Milling and Flour in Ireland', 5.

17 Rynne, C. (1990) op. cit., 23; Monk, M. A. (1981) 'Post-Roman Drying Kilns and the Problem of Function', 216-7; Ó Corráin, D. op. cit., 51.

References

18 Edwards, N. op. cit., 62; Bowie, G. op. cit., 6-7; Monk, M. A. op. cit., 216-8 and 227-8; Power, D. op. cit., 356; Rynne, C. (1990) op. cit.

19 Monk, M. A. op. cit., 227-8; Edwards, N. op. cit., 63; Bowie, G. op. cit., 6.

CHAPTER 6

GRAIN MILLING FROM THE SEVENTEENTH TO THE TWENTIETH CENTURY

The importance of grain growing in seventeenth and early eighteenth century Ireland is reflected in the fact that grain milling was one of the most widespread industries in this period. The main function of the numerous mills dotted throughout the countryside was the grinding on commission of all types of grain for a local market, providing meal for human consumption along with animal feed for the farm. From the middle of the eighteenth century, major changes occurred in Irish milling, with the industry becoming commercialized and large-scale. This important period of expansion continued until the latter half of the nineteenth century, after which the industry went into decline and then consolidated during the twentieth century.[1]

GROWTH OF THE GRAIN MILLING INDUSTRY

In the seventeenth and early eighteenth centuries, vertical-wheeled water-powered mills were predominant, and had displaced most of the smaller horizontal-wheeled mills still in existence at the beginning of this period. Breastshot and overshot wheels were common by the middle of the eighteenth century. The mill itself consisted of two floors, the upper one containing the hopper and one pair of grinding stones and the lower one holding the wheel shaft and gearing. The mills of this period were still quite small and served mainly a local market.[2]

From the middle of the eighteenth century, grain milling in Ireland experienced major change. Large mills for grinding wheat were being established in England. They produced flour that was of a better quality and was more competitively priced than Irish flour at this time, with the result that imports of English flour into Ireland increased. These new mills employed new technology and expanded the milling process. The small mills ground grain without cleaning it. The new mills, however, undertook the shelling and sifting of the grain before grinding and then the mechanical bolting of it afterwards. The former processes removed impurities, while the latter separated the bran and pollards from the flour. Bolting had previously been carried out in a more primitive and expensive way by the bakers. It was an important aspect of the new milling technology and the new mills were generally referred to as 'bolting mills'. These new mills catered for a large British urban market and an export trade to Ireland, and thus required large storage areas for grain and flour. This, along with the expanded milling process, meant that the new mills were much larger structures, of between three and five storeys high.[3] Another distinctive feature of these mills was a much greater use of waterpower. The waterwheel not only drove the millstones, to which it had been confined previously, but now also the shelling, sifting and bolting machinery along with hoists and elevators.[4]

An incentive to establish large mills in Ireland, similar to those in England, was provided by bounties granted by the Irish parliament in 1758 on the inland transport of grain and flour to Dublin. This city was the first area in the country where the use of flour, rather than wheaten meal, was widespread. The bounties had a positive effect on the milling industry but there were other influences too which saw the Irish industry change radically from the middle of the eighteenth century. Irish interest in new methods of industrial organization evident in England was growing, and merchants and landowners were willing to invest readily available capital in the industry. Mill size and output began to increase dramatically. The first large modern mills were built in the counties that were also amongst the most agriculturally advanced regions in the country: Meath and Kilkenny. The greatest of these mills was that erected at Slane, completed in 1767 and operated by David Jebb. From the early 1770s onwards, the construction of large mills, often five or even six storeys high, spread throughout Ireland (Pls. 6.1 and 6.2). As well, some of the old grist mills were modernized and the new technology of the flour mills was adapted by other grain mills for the processing of oats. Much of the construction and installation of machinery in the new mills was carried out by English millwrights and artisans. These new mills were no longer confined to grinding on commission for a very local market, but instead bought, stored, kiln-dried, marketed and sold the grain to growing and often far away markets. Thus, the milling industry became large-scale and commercial. It was the first

Pl. 6.1 Bealick Mill, near Macroom, County Cork. Built in 1818 on the site of an earlier mill, this mill is currently being restored.

Pl. 6.2 Bellmount Mills, Crookstown, County Cork. Built in 1810, this photograph was taken in 1911 and includes the owner, J. P. Howard, in the foreground. (Photograph courtesy of Noel Howard).

industry in Ireland to develop new patterns of industrial buildings and economic organization, and was to be imitated by other industries in the latter part of the eighteenth century.[5]

Continued growth in flour production and in mill construction were features of the milling industry in the late eighteenth and early nineteenth centuries. Though the bounty ended in 1797, other factors contributed to the expansion of the industry. Foster's Corn Law of 1784 increased the export bounties payable on corn, meal and flour, and provided a important boost to the industry. Probably more significant was the impact of the Napoleonic Wars. During these years, 1793 to 1815, Britain was cut off from grain supplies from mainland Europe by a blockade, and was increasingly dependent on flour, meal and grain supplies from Ireland. This was a thriving period for Irish milling. Very large mills, capable of handling up to 40,000 barrels of flour annually, were built. Mill complexes became prominent features of the major grain-growing regions. These included Clonmel which was the largest milling centre in Ireland in this period, as well as the much smaller Nenagh and Ballymahon. Grain and flour prices fell after 1815 with the ending of the war. However, Ireland continued to meet the demands of the expanding British market, as well as the growing home market for flour. Exports of flour grew from 25,000 hundredweight annually in the early part of the nineteenth century to more than 750,000 hundredweight in the years immediately preceding the Famine.[6]

The growth in demand for Irish grain products encouraged investment in steam navigation. The journeytime to England in sailing vessels was so long that flour often went out of condition. The introduction of steam transport had by the late 1820s reduced the time of passage dramatically. Along with flour, British demand for oatmeal was growing. The development of steam navigation had the effect of bringing down the cost of transport on the old sailing vessels. This resulted in an increased quantity of oatmeal being transported by sailing vessel, since this product did not go out of condition quickly.[7] Oats, along with flour, was a major export item in the pre-Famine period, while barely and rye were not important in this trade.[8]

All of these factors together contributed to the expansion of the Irish milling industry in the pre-Famine period and to its continued strength until after the Famine. The latter is indicated by the increase in the numbers of millers in Ireland in this period - from 4,309 in 1841 to 4,417 in 1861. The position of the milling industry in these decades was especially remarkable in the context of considerable deindustrialization in the pre-Famine Irish economy.[9]

Technological Developments

The majority of grain mills in eighteenth and nineteenth century Ireland were water-powered. An adequate supply of water was essential for the efficient operation of the mill. This was provided ideally by a natural reservoir such as a lake, or alternatively by a dam or mill pond which was created by constructing a weir across the course of a river. The water was led from the river to a dam close to the mill via a head race or channel which was cut along the valley side. It was directed from the dam through sluice gates to the waterwheel or turbine (Pl. 6.3). Some smaller mills did not have a dam and the head race passed directly into the mill. Having completed its work, the water was discharged via a tail race into the same river at a point downstream.[10]

Pl. 6.3 Head race, sluice gates, and waterwheel at Bealick Mill, Macroom, County Cork.

Until the latter part of the eighteenth century, the majority of Irish waterwheels were made of timber. These wheels revolved slowly and wore quickly. These problems were rectified around 1780 when the British engineer, John Smeaton, introduced cast-iron shafts, buckets and shrouding. A further very important development, generally known as the suspension wheel, was brought about by other engineers such as T. C. Hewes, William Fairbairn, John Rennie, Robert Casement and Bryan Donkin. They introduced the spur gear rim or segment and pinion or bull wheel, whereby the drive was taken off the rim of the wheel rather than off the axle-shaft, as was previously the case. This had the effect of increasing the power being transmitted from the wheel to the machinery within the mill. It also meant that waterwheels could be built in a free-standing position. This advance allowed for the development of large iron breastshot and overshot waterwheels in the first half of the nineteenth century (Pl. 6.4).[11]

Pl. 6.4 Restored high breastshot, Fairbairn-type suspension wheel with ventilated buckets at Bealick Mill, Macroom, County Cork. This wheel was built before 1860 by McSwiney's of Cork, and was capable of turning up to six pairs of millstones as well as all of the ancillary machinery.

From the 1830s onwards, some of the larger mills invested in steam engines, since even in Ireland water supply during the summer months was often inadequate. However, most of these mills used steam only as a backup to water power. At mid-century, the majority of grain mills were waterpowered. The preference for water power was linked to the improved design and efficiency of waterwheels in the first half of the nineteenth century, as well as the lower cost of water power in the winter months during the very busy milling season.[12]

Along with advances in waterwheel manufacture, the availability of large quantities of inexpensive cast iron furthered the development of the internal mill machinery. Its use was also encouraged by the shortage and increasing cost of timber during the Napoleonic Wars. The development of cast iron shafting and gearing, along with the great spur wheel, meant that power from the waterwheel could be transmitted to more than one set of millstones, and to ancillary machinery such as hoists and elevators (Fig. 6.1).[13]

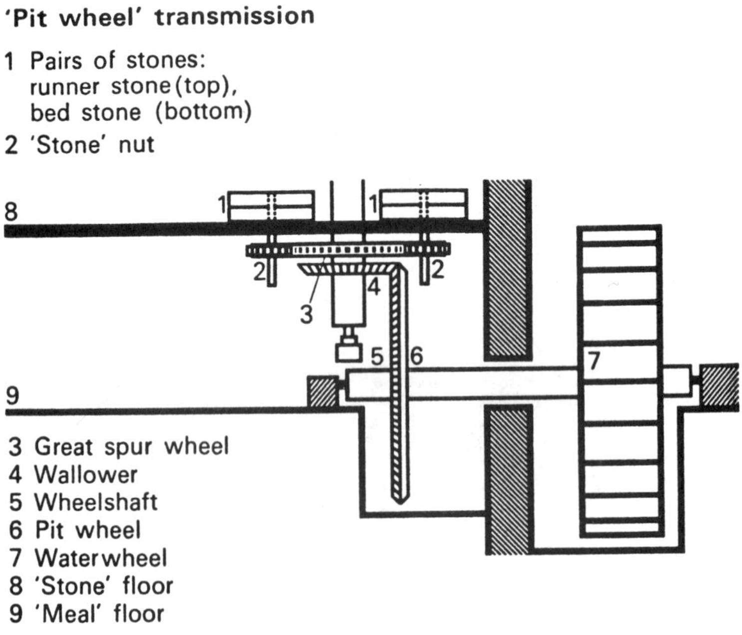

Fig. 6.1 Water-powered mill gearing and power transmission (After McCutcheon, W. A., 1980, p. 236).

Up to fifteen pairs of millstones could be operated simultaneously in many of the new grain mills. In this period, millstone diameters generally ranged from three and a half feet to four and a half feet. Millstones had to be constructed from carefully chosen stone to meet exacting requirements.[14] The principal types of natural stone used in grain milling were a crystalline granodiorite, or a hard primary limestone or grit. Stone was supplied to the Irish milling industry by a number of quarries in the country, particularly those at Hook Head in County Wexford, Ballyhoura Hill in County Cork and Innish-na-mBro on the Blasket Islands, off the Kerry coast. A tough millstone grit stone which was particularly suited to grain milling and which was used in Ireland was 'Peak Stone', quarried in the Peak District of Derbyshire, England. Of all stone types, the preferred one for flour milling was French 'burr' or quartzite (Pl. 6.5). Quarried in the Paris Basin, it was the hardest and finest type of stone available. The millstone was usually constructed from wedged-shaped pieces of this rock, which were cemented together with plaster of Paris and bound with an iron band shrunk on round the rim.[15]

Pl. 6.5 Dressed French 'burr' millstone at Clonmoyle Mill, County Cork (Photograph courtesy of Noel Howard).

The grinding faces of both the upper runnerstone and the lower bedstone were cut with furrows or grooves which radiated from the centre. A scissors-like cutting action was produced when the stone surfaces passed over one another. In order to keep the cutting edges sharp, the grinding surfaces of both stones had to be frequently dressed or picked. This was a skilled job carried out by the miller or occasionally by a journeyman stone dresser, and required the use of special tools such as chisels, picks and mill bills.[16]

The milling process of the new large mills included not only grinding - the function of the small mills - but also shelling, sifting and bolting. The top floor of these multi-storeyed mills was a storage area, the grain being hoisted there by mechanical means. On this floor, the grain was loaded into bins from which it was directed through gravity chutes to the hopper in the middle floor. The millstones, placed in pairs, were located in the latter floor. The grain was fed into the eye of the runnerstone by means of a 'shoe' or short chute attached to the hopper, and was shelled. The next stages involved the sifting and grinding of the flour, again in the middle floor or floors. Following grinding, the grain was fed into a bolter. This consisted of a cloth or wire sleeve, through which the fine flour passed and was separated from the bran and pollards which were retained. The flour was then ground a second time, and was directed to the ground floor where it was bagged. This floor also contained the shafting and gearing through which power from the wheel or other source was transmitted to the millstones and other machinery above. The processing of oats also benefited from the adaptation of this new technology, with fans, sieves and dusters performing the same function as the bolters in the flour mills. These new processes gradually dispensed with the practice of going to high windy locations, often known as 'shelling hills', to hand winnow shelled oats.[17]

Along with the above technological changes experienced by the milling industry from the late eighteenth century, the methods by which grain was dried and stored were also transformed. The new merchant millers took responsibility for the kiln-drying of the grain. The traditional free-standing kilns were replaced by 'standard flat-head' kilns, which were either incorporated within the mill building or were independent structures close to the mill (Fig. 6.2). These kilns usually consisted of two storeys. Heat was provided by a fire which burned in a 'kiln pot' on the ground floor.

Peat and oat hulls or 'seeds' were the sources of fuel for the fire, particularly in the poorer areas of the north and west of Ireland. In the more fertile tillage regions of the southern half of the country, the use of anthracite, or culm, was common. This produced great heat which was necessary for the rapid drying of wheat and barley, and produced little smoke which meant that the taste of the grain was not adversely affected. The second storey contained the 'kiln head' which had a floor of perforated metal sheets or tiles, or wire mesh. The smoke and heat wafted up through this floor and dried the grain which was spread on it to a depth of up to six inches and which was turned over regularly by hand. The smoke was emitted though a vent in the roof.[18] These kilns remained in use for two centuries.

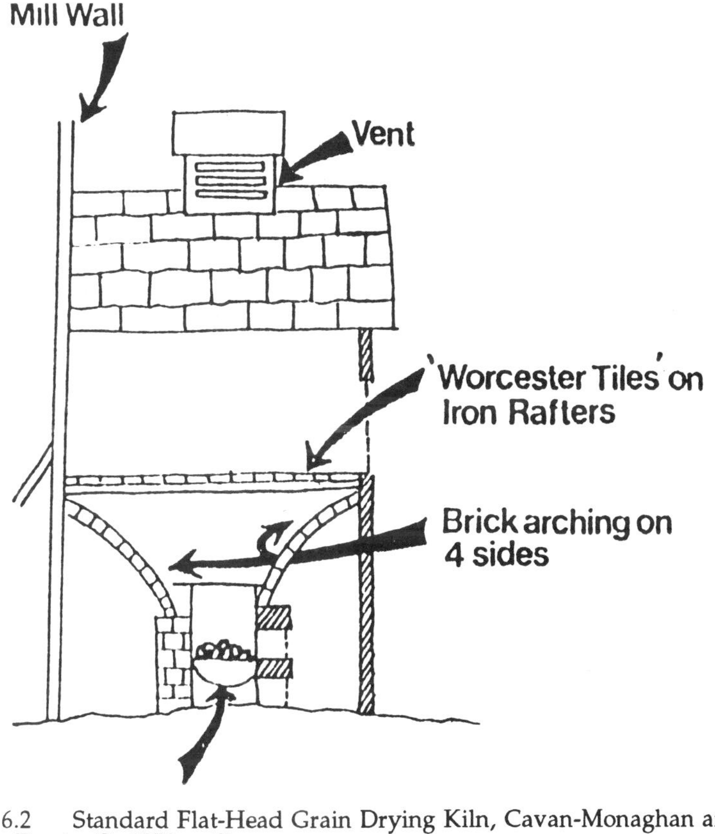

Fig. 6.2 Standard Flat-Head Grain Drying Kiln, Cavan-Monaghan area (After Bowie, G., 1979, p. 8).

With the growth in grain production from the late eighteenth century, large grain stores and granaries became a feature of the Irish landscape. Most of these buildings were erected around the coastline or along navigable waterways during or after the Napoleonic Wars when grain exports were at a peak (Pl. 6.6). These stores were usually between five and seven storeys high. Initially, the bags of grain were taken up through these floors manually but later hoists and other mechanical devices were introduced. Some drying of the grain was carried on in these buildings, by the simple process of allowing fresh air through the windows and by frequent turning of the grain.[19]

Pl. 6.6 Grain store at The Lee Mills, Cork (rebuilt 1825-31).

DECLINE AND CHANGE IN THE GRAIN MILLING INDUSTRY

The Irish milling industry began to experience dramatic change in the second half of the nineteenth century, particularly from the 1870s onwards. A drop in the price of wheat on the international market brought about a severe reduction in wheat acreage in Ireland, from 64,901 acres in 1851 to 6,038 acres in 1901. This in turn had a negative effect on the milling industry. The Corn Laws were repealed in 1846, with the result that the British market was opened up to cheaper supplies from continental Europe and increasingly from North America. Also, it is likely that the post-Famine population decrease in Ireland caused a contraction in the domestic flour market. Exports of flour from Ireland to England dropped to 160,000 hundredweight during the 1860s and even more drastically from the late 1870s. This was brought about mainly by the importation of cheaper American flour into Britain. By this time, new technology was being introduced into the milling industry and was to change it radically. This new technology was quickly adopted in North America. Iron rollers which could grind larger quantities of grain in a much shorter time and thus more cheaply than the traditional millstones had been developed. This development together with cheaper shipping had the effect of reducing the international price of flour. The impact of these developments on the Irish milling industry was severe, sending it into crisis in the late 1870s and 1880s. Within ten years, two-thirds of Irish mills had closed down and further closures of small country grain and flour mills using millstone technology had occurred by the end of the century. Some mills in the inland areas did modernize but for most the reduction in wheat acreage meant that this option was not economically viable. The mill complexes in the interior, including Clonmel, went into decline. The situation was different for the mills in the port towns. Ports such as Belfast, Cork, Limerick and Dublin adopted the roller milling technology and large-scale enterprise became a dominant feature of the milling industry. These ports were in a position to grind cheap, imported American wheat into flour for either the home market or for re-export to England, or to apply the new technology to grinding Irish wheat into flour and imported maize into animal feed. Cork was one of the ports to reorganize and had become a major flour milling centre by the early 1890s while Limerick and Dublin had the largest single mills in Ireland. For example, the milling firm of J. N. Russell & Sons built a roller-mill at Newtownpery in Limerick which was nine storeys high, was powered by a 350-horsepower steam engine and boasted nineteen sets of rollers for wheat. This was the largest individual mill in the country at the end of the nineteenth century.[20]

There was some resurgence of the milling industry in the early decades of the twentieth century. During the 1930s, the Fianna Fáil government offered a guaranteed price for native wheat. Millers were obliged to use a stated amount of Irish wheat in their flour and strict controls were imposed on flour imports. The greater use of native grains, for example barley, in animal feed was also promoted by restrictions on maize imports. During both world wars, Ireland faced serious shortages of wheat and flour. Grain acreage increased considerably in these years, encouraged during World War II by the government introduction of compulsory tillage. These factors together contributed at least temporarily to the revival of the industry and significantly of country milling.[21]

The trend toward large-scale enterprise intensified in the Irish milling industry in the postwar decades. Today, flour milling is controlled by four large milling groups - Odlums (Greencore Group) and Bolands (I.A.W.S. Group) in the Republic who own six milling plants and the Andrews Group and Allied Mills in Northern Ireland who control three mills. Some of these mills, such as Howard Bros. in Crookstown, County Cork, continue to stone grind their products, and the renewed interest in stone-ground wholemeal flour is encouraging the continuation of this tradition (Pl. 6.7).[22]

Pl. 6.7 Cork Summer Show, 1947: Howard's "Oneway" flour and "Thrivo" foods (Photograph courtesy of Noel Howard).

A reduction in the acreage of oats from the end of World War II brought about a major decline in the number of rural waterpowered grain mills producing oatmeal. Once a major part of the staple rural diet, oatmeal is now ground in just a few mills.[23] Today, oats are mainly grown for animal feed, with only ten to fifteen per cent being used to make oatmeal. Some of the country mills did survive after the 1940s, mainly milling oats, barley and maize for animal feed. This was partly linked to an increase in calf, pig and poultry rearing directly after the war. With the development of diesel power, rural milling in the postwar decades was not centred around the waterpowered mills but rather around mills which were associated with and sometimes operated by the creameries. From the 1960s onward, however, most of the rural mills went out of business. The main reasons for this was a decline in pig and poultry rearing and other changes in farming methods, along with the growth of large provender mills providing bulk supply facilities which in turn went through a series of amalgamations. Consequently, the milling of animal feed, along with products for human consumption, is today controlled by a few large groups. [24]

Technological Changes

In the late nineteenth and early twentieth centuries, the milling industry experienced a number of technological changes, many of which were responsible for its spatial contraction and for the development of large-scale enterprises. The introduction of roller milling technology had the greatest impact. Unlike the stone milling process which ground wheat into bran and flour in one operation, this new system gradually reduced and separated the various parts of the wheat at different stages. The wheat was passed through the first set of rollers which opened the husk and sifted the middlings or floury part from the bran. It then went through further pairs of rollers, being sifted each time to remove any remaining coarse material. Along with being more efficient, this new system produced a greater proportion of fine white flour for which there was a growing taste, unlike millstones which produced coarser grades.[25]

Some mills, however, continued to use millstones. French burr stones were in common use until the early twentieth century. From that period, 'composition' stones made up of emery powder or ground French burr and magnesite cement were being developed and now are predominant among all millstone types. As well as being less expensive than the burr stones, these composite stones are less difficult to dress. The traditional tools and skills of the stone dresser have been replaced by the use of electric powered angle grinders.[26]

Along with changes in internal milling machinery, developments in the type of power used also occurred. Improvements in the waterwheel in the pre-Famine period led to the development of the water turbine from the middle of the nineteenth century. These were compact wheels which were fully enclosed in a casing. Cups situated on the wheel's circumference were rotated within this casing at high speed by water directed through a nozzle. These powerful turbines replaced waterwheels in a number of grain mills throughout the country, and some are still used to generate DC electricity supply for domestic and farm use.[27] However, turbines did not become predominant as prime movers in Irish grain mills. This was due to the increased use of steam power as the milling industry became more large-scale and concentrated on the coastline in the latter part of the nineteenth century. These large mills required considerable power, and steam engines could be produced in several hundred horsepower. As well, the fuel source for these engines, coal, was readily available at the ports. Steam remained an important source of power for the large mills up to the 1950s. From the 1920s, the remaining smaller country mills began to increasingly rely on diesel engines such as Crossley, Blackstone and National, though waterpowered mills temporarily regained their importance when diesel supplies became restricted during World War II. From the 1950s, the majority of mills in Ireland had switched to the use of large diesel engines or electric motors.[28]

Grain drying methods remained unaffected by the post-Famine changes in milling. The standard flat-head kilns continued to be used and it was not until the middle of the twentieth century that change came about. Unlike the more traditional harvesting methods which allowed for some drying of the grain crop before threshing, the introduction of the combine harvester in the 1950s meant that the grain was harvested from growing straw. This resulted in an increasing requirement for drying facilities. This demand was met by the introduction of oiled-fired forced hot-air driers.[29]

Grain storage methods experienced change more rapidly. Many of the large stores and granaries which had been built from the latter decades of the eighteenth century went out of use with the general decline of the milling industry a century later. While some did survive and are still in use, grain is more usually stored today in large circular metal constructions, known as 'silos'. These are found close to the large provender and flour mills.[30]

WIND-POWERED GRAIN MILLING

Wind power was used in Ireland for grain milling, though it was never a prime mover. The earliest vertical windmills in western Europe were post mills and came into use in medieval times. This type consisted of a central pivotal post upon which the sails machinery within the mill rotated. One of the earliest known windmills in Ireland was a post mill, erected near Old Ross, County Wexford, around 1305. However, this type did not become widespread in Ireland.[31]

The majority of Irish windmills were those known as tower mills, and these were common in the eighteenth and early nineteenth centuries (Pl. 6.8). These were fixed circular stone towers with a moveable cap resting on top. The windshaft carrying the sails was contained in the cap. The remainder of the machinery was located throughout the four or five floors of the tower. The milling process was vertical and in this respect was similar to waterpowered mills. The tower mills benefited from a number of mechanical improvements which were introduced during the eighteenth and nineteenth centuries. Of particular importance was the 'fantail', invented in England by Edmund Lee in 1745. This was a device which automatically kept the sails facing the wind.[32]

Windmills in Ireland were overwhelmingly concentrated in counties Down and Wexford. There were a number of reasons for this. The sheltered nature of this eastern coastline, and its drier and warmer weather encouraged widespread cereal cultivation in these counties. Both were close by sea to Dublin, which provided a large urban market for grain products. The drumlin topography of County Down was such that windmills and watermills could be used in conjunction with one another. In the 1840s, for example, County Down had over one hundred windmills, most producing oatmeal and flour. By the end of the century, few were working. The number of windmills now surviving in Ireland is very small. However, four have been restored: Elphin, County Roscommon, built in the early eighteenth century; Ballycopeland, County Down, erected in the 1780s; Tacumshin, County Wexford, constructed in the mid nineteenth century; and Blennerville, County Kerry, completed in 1800. The latter is the largest working windmill in the British Isles (Pl. 6.9).[33]

Pl. 6.8 A restored early eighteenth century tower mill at Elphin, County Roscommon. In order to face the sails to the wind, the cap was turned by pushing the 'cart-wheel' device.

Pl. 6.9 Blennerville Windmill, Tralee, County Kerry.

The middle of the eighteenth century marked a major period of change in Irish grain milling. The transformation of milling into a large-scale, highly commercialized industry was attributable to a number of factors, including the 1758 government bounties on the inland carriage of grain, Foster's Corn Law, and the impact of the Napoleonic Wars. Technological advances, the extension of the milling process and the provision of kiln-drying facilities at the mill, were features of this transformation. The establishment of large grain stores and granaries paralleled the developments in milling. The period of decline and change which followed this dramatic expansion was brought about by a combination of factors, among them a drop in the international price of flour, repeal of the Corn Laws, and the importation of cheap American flour into Britain. Some mills, particularly in the ports, survived by reorganizing and introducing new technology, particularly roller milling. Grain storage methods changed also, as did drying methods to a slower extent. The survival of the Irish milling industry in the twentieth century was achieved to a degree by government policy in the first few decades, and significantly by the consolidation of the industry into a few large milling groups in more recent years.

References

1 Cullen, L. M. (1977) 'Eighteenth-Century Flour Milling in Ireland', 5; Shaffrey, P. and M. (1985) *Irish Countryside Buildings*, 91.

2 Cullen, L. M. (1977) op. cit., 6-7; McCutcheon, W. A. (1980) *The Industrial Archaeology of Northern Ireland*, 232.

3 Cullen, L. M. (1977) op. cit., 7-9 and 16.

4 Bielenberg, A. (1991) *Cork's Industrial Revolution 1780-1880*, 41.

5 Cullen, L. M. (1977) op. cit., 9-10 and 18-22; Cullen, L. M. (1987) *An Economic History of Ireland since 1660*, 77 and 92-4; Bielenberg, A. op. cit.; Bowie, G. (1979) 'Corn Drying Kilns, Meal Milling and Flour in Ireland', 9-11.

6 Bielenberg, A. op. cit., 42; Cullen, L. M. (1987) op. cit., 96, 100-1 and 146-7; Ó Gráda, C. (1994) *Ireland: A New Economic History, 1780-1939*, 305.

7 Bielenberg, A. op. cit., 44 and 120.

8 Bourke, P. M. A. (1976) ' The Irish Grain Trade, 1839-48', 162-3.

9 Ó Gráda, C. (1988) *Ireland before and after the Famine*, 31; Cullen, L. M. (1987) op. cit., 146.

10 McCutcheon, W. A. op. cit., 235.

11 McCutcheon, W. A. op. cit., 234; Buchanan, R. A. (1982) *Industrial Archaeology in Britain*, 237 and 242; O'Sullivan, D. J. (1995) *The Story of Corn and Flour Milling in Ireland*, 10.

12 Rynne, C. (1993) *The Archaeology of Cork City and Harbour*, 87; Bielenberg, A. op. cit., 46.

13 McCutcheon, W. A. op. cit., 236-7; Bielenberg, A. op. cit., 43.

14 Cullen, L. M. (1987) op. cit., 92; Rynne, C. op. cit.; Tucker, D. G. (1987) 'Millstone making in England', 167

15 McCutcheon, W. A. op. cit., 238; O'Sullivan, D. J. op. cit., 19; Buchanan, R. A. op. cit., 196-7.

16 McCutcheon, W. A. op. cit.; O'Sullivan, D. J. op. cit., 21.

17 Buchanan, R. A. op. cit., 195-6; Cullen, L. M. (1977) op. cit., 8-9; McCutcheon, W. A. op. cit., 238-9 and 275; Bowie, G. op. cit., 9.

18 Bowie, G. op. cit., 7; McCutcheon, W. A. op. cit., 239; O'Sullivan, D. J. op. cit., 25.

19 O'Sullivan, D. J. op. cit., 26; Power, D. et al (1992) *Archaeological Inventory of County Cork*, 364.

20 Bielenberg, A. op. cit., 46-9; Ó Gráda, C. (1994) op. cit.; Cullen, L. M. (1987) op. cit., 147 and 156-7; Green, E. R. R. (1969) 'Industrial Decline in the Nineteenth Century', 98; Cullen, L. M. (1977) op. cit., 25.

21 Daly, M. E. (1981) *Social and Economic History of Ireland since 1800*, 147; Bielenberg, A. op. cit., 48; O'Sullivan, D. J. op. cit., 39.

22 O'Sullivan, D. J. op. cit., 32-3; Tucker, D. G. op. cit., 168.

References

23 Bowie, G. op. cit., 9.
24 O'Sullivan, D. J. op. cit., 23-4, 28, and 35
25 Perrin, R. (1990) 'Structural change and market growth in the food industry: flour milling in Britain, Europe, and America, 1850-1914', 423-4; Department of Agriculture and Technical Instruction for Ireland (1902) *Ireland - Industrial and Agricultural*, 402.
26 Tucker, D. G. op. cit., 168; O'Sullivan, D. J. op. cit., 20-1.
27 Buchanan, R. A. op. cit., 242-3; McCutcheon, W. A. op. cit., 260-3.
28 O'Sullivan, D. J. op. cit., 14-5, 23 and 39.
29 Ibid, 27; Bowie, G. op. cit., 7.
30 O'Sullivan, D. J. op. cit., 26-7; Power, D. et al, op. cit.
31 O'Sullivan, D. J. op. cit., 29; O'Sullivan, A. M. (Undated) *Tacumshin Windmill - its history and mode of operation*, 1.
32 McCutcheon, W. A. op. cit., 228-9; Buchanan, R. A. op. cit., 232.
33 McCutcheon, W. A. op. cit., 229; O'Sullivan, A. M. op. cit., 2; O'Sullivan, D. J. op. cit., 30.

CHAPTER 7

GRAIN PRODUCTS AND THEIR USES

The process involved in converting grain into useful products was a lengthy one. Following the preparation of the soil and the growing period, the ripe crop went through the stages of harvesting, threshing, winnowing, drying and then milling, or brewing and distilling.[1] Grain is a versatile crop from which a variety of products have been gleaned. An essential part of the human diet, grain has been used particularly to make bread and porridge. It has also been used in the production of alcoholic drink for human consumption. Grain has been the primary source for animal feed and for animal bedding. Other uses have included roof thatch, straw ropes and objects associated with festive occasions.

GRAIN IN THE IRISH DIET

One of the most important uses for grain in Ireland has been in the production of food for human consumption. From Neolithic times up to the seventeenth century, grain along with milk were the staple elements of the Irish diet. Evidence for the importance of grain in early times is provided by the numerous archaeological discoveries of querns, watermills and grain-drying kilns, as well as the remains of cereal crops.[2] Early Irish literature provides further evidence. Between the early sixth and the early ninth centuries, the annals record on five occasions the destruction of the grain crop as a result of bad weather. Significantly, famine resulted at least twice. The uprooting or burning of the enemy's grain is also recorded by the annals. This was a common military tactic in the eleventh and twelfth centuries and it highlights the value put on the grain crop. The poem *Vision of MacConglinne* is an excellent record of the Irish diet in the twelfth century. Oats, barley, wheat and rye are among the foods listed, and in medieval times their importance was in that order. These cereals were used to make porridge and bread. While the latter was important in the diet in that period and in subsequent centuries, porridge and gruel were the more common elements.[3]

Irish diet began to change from the seventeenth century onwards. Major growth in population and trade associated with low income levels brought about significant commercialization of food surpluses. This occurred particularly in the butter trade. With a reduction in the availability of dairy products for home consumption, particularly among the poor, reliance on grain and potatoes grew. The potato was used with grain in the winter diet, while the latter became the main food in the spring. The diet was becoming more varied in this period. With regard to grain consumption, the emphasis changed from the relatively simple food of porridge to bread made of various cereals.[4] While consumption of the potato grew during the eighteenth century, especially among the poorer sections of society, it did not dominate the Irish diet. Grain continued to be important in the rural and urban diet. The consumption of oaten bread was widespread, and the population of Dublin, in particular, created a huge demand for wheaten bread. Oatmeal was essential in the diet of the poor through the summer months. It was only as population and unemployment increased in the late eighteenth and early nineteenth centuries that the potato became the predominant element of the diet. However, there were regional and class variations. The potato was particularly associated with cottiers, labourers and smallholders in the poorer regions of the country.[5] Even among these groups, cheap imported Indian meal or maize meal was being used from as early as 1800 in the early summer or 'hungry months' when the potato was not available.[6] As well, in the north and the east of the country, wage-paid labourers and others involved in non-agricultural employment had enough cash resources to buy marketed products, of which oatmeal and bread were the more significant.[7] Farmers throughout the country continued to enjoy a mixed diet of oatmeal, wheaten bread, potatoes and milk. Even in the quarter century before the Famine, two-thirds of the large quantities of grain grown in Ireland were consumed in the country.[8]

The Great Famine had a catastrophic effect on the three million people who relied almost entirely on the potato for food and in this regard it was regionally and socially selective.[9] Its impact on those whose diet included grain was not so severe. The role of grain in the Irish diet during the Great Famine has been the subject of much controversy. The populist view is that the major shortfall in food supply with the failure of the potato crop in the latter half of the 1840s could have been made up if a total ban on all grain exports from Ireland had been implemented. In fact, Ireland did move from being a large exporter of grain to a net importer during the Famine. As well, the quantity of grain being produced in the country at this time would not have been sufficient to meet the food deficiency. However, a total prohibition on exports in 1846 and 1847 would have saved lives during the period of delay which occurred in obtaining foreign maize supplies. On a political level, the removal of relief entitlements from people on holdings of over one quarter of an acre of land added substantially to the numbers who died during the Famine.[10]

The impact of the Famine on diet was not as great as is usually believed. Grain, in the form of imported maize, increased in importance in the diet of the poorer families and regions. However, it was not until the 1870s and 1880s that diet began to change generally. People in rural areas began to adopt urban diets, and this was part of a widespread revolution in Europe. A shift away from home-made products such as oaten bread to white bakers' bread was particularly significant.[11] This was a reflection of an increasing reliance on shops for food and other products[12], and it marked the beginning of a growing variation in the Irish diet. This change in diet has become particularly evident over the past number of decades. Grain continues to be important in the Irish diet today and now in its many forms is the basic ingredient of numerous foods.

In the past, porridge and bread were the principal foods made from grain. Porridge or stirabout was usually made from oats, but also from wheat, barley and for a number of decades after the Famine from Indian meal. Oatmeal stirabout, made with milk or water, was the main breakfast food in rural and urban areas well into the nineteenth century, while in earlier times porridge was often the main element of all meals.[13] Porridge was made in a variety of ways. From at least the ninth century, a mixture of raw oatmeal and butter, known as 'menadach', was eaten. A similar dish, but with thick milk, cream or buttermilk instead of butter, and sometimes hardened wheat instead of oatmeal, was eaten until recently in some areas, including Cape Clear off the south coast. This dish was known by such names as 'praípín' or 'ríobún'. Fermented oats husks were used to make porridge and more commonly to make a popular drink, both known as 'sowans'. Porridge continues to be important in the diet today, despite competition from modern breakfast cereals. Traditional stoneground oatmeal is still produced at just a few mills, including the distinctively-flavoured 'Macroom Oatmeal', ground at Walton's Mill, Macroom, County Cork (Pl. 7.1). Flahavans of County Waterford and Odlums of Dublin are currently the largest manufacturers of oatmeal in Ireland.[14]

Pl. 7.1 A 'Macroom Oatmeal' packet from the 1920s. This oatmeal was famous throughout counties Cork and Kerry for its taste and flavour (Packet courtesy of Donal and Teresa Creedon).

From early times, the most common grain used for making bread was oats. Wheaten bread was considered a luxury. Barley bread was regarded as being inferior to both and in medieval times was eaten for the purpose of mortification by monks and others. Rye was rarely used for bread-making, though such bread was a Christmas treat in Connemara. Bread was sometimes made from mixed cereals, for example, from oats and wheat, maize meal and wheat, and also from flour and potatoes. Oat bread was very thin and was usually baked without leaven.[15] It was common in Ireland well into the nineteenth century[16], but by this time wheat bread was becoming more commonplace in the diet. The tradition of home-made country soda bread began in the early nineteenth century with the use of whole wheat meal or 'brown flour' together with soda which increasingly replaced yeast, barm and sour dough as a leavening agent. However, it was the greater use of bleached wheaten flour or 'white flour' in bread-making which was responsible, to a great extent, for a revolution in the ordinary diet in the second half of the nineteenth century. White bread, often made from cheap imported grain or flour, became fashionable in this period, though it was less nourishing than brown bread. With the use of white flour came the growth of town bakeries. White bakery bread was considered to be superior to that baked at home and its introduction resulted in a decline in the importance not only of oaten bread but also of brown bread in the Irish diet. Bakers' bread, along with a variety of cakes, were bought directly from the rural horse-drawn bread vans which were very common by the end of the nineteenth century or from local shops which were supplied by them.[17]

Home-made bread was baked by a variety of methods in Ireland. Wall-ovens were common in the houses of well-off families particularly in the east and south of the country but were unknown amongst the majority of the population in Ireland. In early times, bread was baked by placing it directly on the coals of the open fire, or on a large cabbage leaf or heated flagstone. This practice was superseded by the baking of bread in front of the fire with the support of 'hardening stands' or fenders, which were made of timber, stone or wrought-iron (Fig. 7.1). In more recent times, bread was baked either on a griddle or in a pot oven known as a 'bastable'. The griddle was generally a circular flat iron plate with two handles (Pl. 7.2). The pot oven was deep and was made of cast iron. It had a lid upon which glowing embers were placed to ensure even baking. Both utensils could be either hung or placed on a trivet over the fire. Bread made in these ways was thin and could be leavened or unleavened.[18]

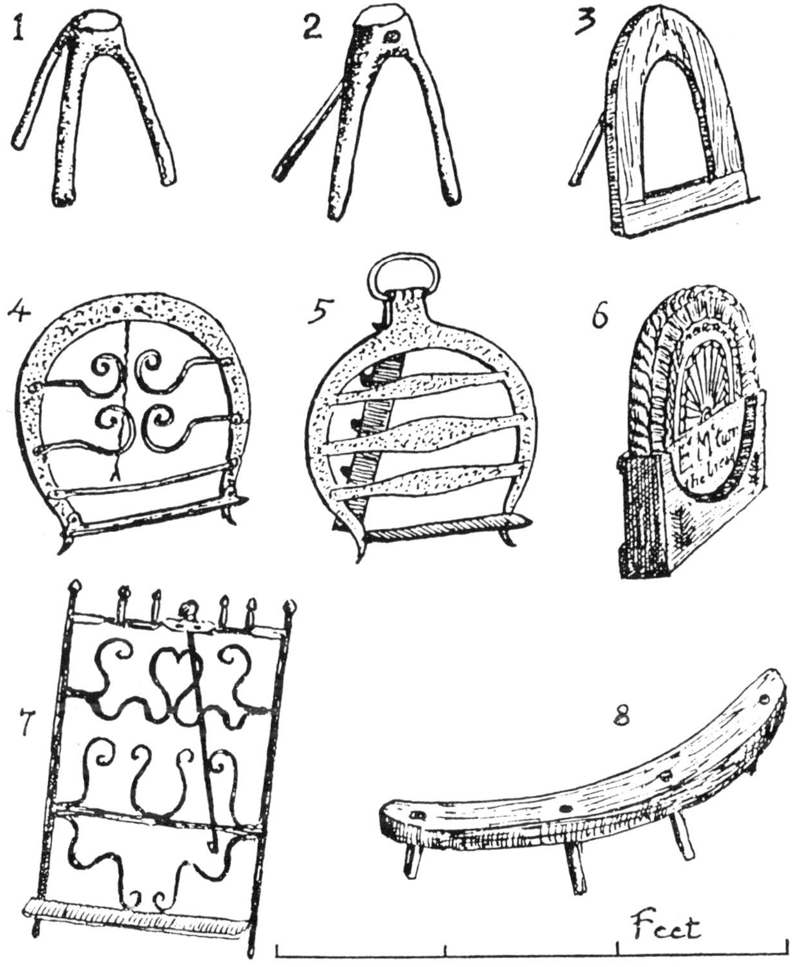

Fig. 7.1 Hardening stands or oatcake toasters: (1), (2) and (3) 'bread-sticks' from Armagh, Galway and Clare, respectively; (4) and (5) bread-irons from Derry and Down, respectively; (6) hardening stand made from stone, Fermanagh; (7) bread-iron from Armagh; (8) toasting fender from Antrim (After Evans, E. E., 1957, p. 80).

Pl. 7.2 Bread being baked on a griddle (Bunratty Folk Park, County Clare).

OTHER USES FOR GRAIN

Along with its importance in the Irish diet, grain has had a number of other uses over the years. Barley has been used for centuries in the brewing and distilling of alcoholic drink. Ale was the most common alcoholic drink in the early medieval period. It was important in the economy, since malt was often used to pay rents and legal debts. Distillation of whiskey began during medieval times.[19] This process became commercialized only in the mid eighteenth century, though much of the supplies continued to come from illicit stills. Whiskey initially replaced the other main drinks at that time - ale, cider, wine and brandy. Despite the blow to the industries by Father Matthew's temperance campaign of the late 1830s, distilling and brewing survived.[20] The latter industry was particularly successful in the following decades, its output trebling between the mid nineteenth century and 1914. This was largely due to the growth of the Dublin brewery, Arthur Guinness, which by 1833 had succeeded Beamish and Crawford in Cork as the principal brewers in Ireland and by 1914 had become the largest brewery in the world. The raw materials for both the brewing and distilling industries - the main one being barley - were to a large extent produced in Ireland. Oats were also used in the distilling industry to produce mixed grain pot-still whiskey. A major use for barley grown in Ireland continues to be in the production of beer, ale, porter, whiskey, and even illicit poteen. Today, malsters, brewers and distillers are supplied with malting barley grown on contract mainly in the eastern and southern counties of Ireland.[21]

Grain has been a very important source of animal feed since early times. Traditionally, oats and oat straw were the main products used to feed cattle, horses and poultry. In the nineteenth century, Indian meal became important, and by the middle of the present century, this grain, along with the native oats and barley, was the main animal feed (Pl. 7.3). From this period, there was a significant expansion in the use of barley for animal feed, which to a substantial degree replaced oats. In more recent years, there has been a growing demand for compounded feedstuffs . A wide variety of feed mixes made from barley, oats, wheat and maize are available today.[22]

Pl. 7.3 Animal feed being loaded for transportation to Waterford at Clondulane railway depot, near Fermoy, County Cork, 1959. This depot was a distribution centre for Clondulane Mill. Flour and other grain products were transported from the mill to the depot, a distance of about half a mile, via overhead steel cables (Photograph courtesy of Jim O'Doherty).

Grain straw has been put to a variety of uses throughout the centuries in Ireland. From early times, roofs were thatched using wheat, oat and barley straw, as well as rye straw which was popular along the west coast. The latter had the longest life, with the roof needing recoating only about every twenty years. Oat straw was one of the most common materials used for animal bedding. Straw, and particularly oat and wheat straw, was widely used to make ropes, known as 'súgán'. Traditionally, these ropes were made by using a 'thraw-hook', a tool made from ash or hazel, and more recently by the use of a twister made of wire. Two people were involved in the construction of a straw rope. One fed the threshed straw into the rope while the second twisted it at the other end using a thraw-hook. Straw rope was used in the construction of granaries, and in a variety of other objects, such as baskets, mats, chairs, mattresses, beehives, horse harness and animal tethers. Straw was also used for festive occasions. St. Brigid's Day was marked by the making of straw or rush crosses in a variety of shapes. Harvest knots, an important element of the harvest festivities, were made from flax, oats or wheat. Straw caps and masks were worn as disguises at Hallowe'en and by Mummers on their visits around the locality at Christmas time.[23]

As a food for human consumption, grain has been significant from early times. Even in pre-Famine times, grain continued to be an element of the Irish diet, despite the importance of the potato in that period. Porridge, home-made bread, and increasingly bakers' bread from the late nineteenth century, were the main foods made from the grain seed. The diversification in diet in the twentieth century has increased considerably the variety of foods containing grain as an ingredient. Grain has been the primary product used in the brewing and distilling industries. It has been an essential element of animal feed, while grain straw has had several uses over the centuries. The wide variety of grain products underlines the importance of this crop in Irish society and economy through time, and accounts for the effort and time that has been in the past, and still is, invested in growing, harvesting, threshing and milling grain in Ireland.

References

1 Ó Danachair, C. (1955) 'The Flail and other Threshing Methods', 6.

2 O'Neill, T. P. (1977) *Life and Tradtion in Rural Ireland*, 56-7; Rynne, C. (1993) *The Archaeology of Cork City and Harbour*, 25 and 49.

3 Ó Corráin, D. (1972) *Ireland before the Normans*, 50-2; O'Neill, T. P. op. cit., 57; Edwards, N. (1990) *The Archaeology of Early Medieval Ireland*, 60; Cullen, L. M. (1981a) 'Population Growth and Diet, 1600-1850', 97.

4 Cullen, L. M. (1981b) *The Emergence of Modern Ireland 1600-1900*, 145; Cullen, L. M. (1981a) op. cit., 99-100 and 111.

5 Cullen, L. M. (1987) *An Economic History of Ireland since1660*, 70, 118 and 130; Dickson, D. (1983) 'The place of Dublin in the Eighteenth-Century Irish Economy', 186.

6 Daly, M. E. (1981) *Social and Economic History of Ireland*, 96; Crawford, E. M. (1981) 'Indian Meal and Pellagra in Nineteenth-Century Ireland', 115 and 118.

7 Mokyr, J. (1981) 'Irish History with the Potato', 13; Cullen, L. M. (1987) op. cit., 130; Clarkson, L. A. and Crawford, E. M. (1988) 'Dietary Directions: A Topographical Survey of Irish Diet, 1836', 175.

8 Cullen, L. M. (1987) op. cit., 118; Bell, J. and Watson, M. (1986) *Irish Farming - Implements and Techniques 1750-1900*, 178; O'Neill, T. P. (1973) 'Poverty in Ireland 1815-45', 30.

9 Cullen, L. M. (1981a) op. cit., 112; Ó Gráda, C. (1988) *Ireland before and after the Famine*, 9.

10 Ó Gráda, C. op. cit., 109-110; Bourke, P. M. A. (1976) 'The Irish Grain Trade, 1839-48', 165.

11 Cullen, L. M. (1981b) op. cit., 146 and 158.

12 Cullen, L. M. (1987) op. cit., 138.

13 Danaher, K. (1985) *The hearth and stool and all! Irish Rural Households*, 60; Daly, M. E. op. cit.; Cullen, L. M. (1981b) op. cit., 175; O'Neill, T. P. (1977) *Life and Tradition in Rural Ireland*, 57-8.

14 Mahon, B. (1980) 'Oatmeal as Food', 123-4; Danaher, K. (1985) op. cit., 66; O'Neill, T. P. (1977) op. cit., 57 and 67; O'Sullivan, D. J. (1995) *The Story of Corn and Flour Milling in Ireland*, 28.

15 O'Neill, T. P. (1977) op. cit., 57 and 64; Danaher, K. (1962) *In Ireland Long Ago*, 46-50

16 Mahon, B. (1980) op. cit., 123.

17 Danaher, K. (1985) op. cit., 58-60; Cullen, L. M. (1981b) op. cit., 183-5; Evans, E. E. (1957) *Irish Folk Ways*, 78.

18 Cullen, L. M. (1981b) op. cit., 183-4; Danaher, K. (1962) op. cit., 46-48; O'Neill, T. P. (1977) op. cit., 58.

19 Ó Corráin, D. op. cit., 66; Edwards, N. op. cit., 62; O'Neill, T. P. (1977) op. cit., 59;

20 Cullen, L. M. (1981b) op. cit., 172-3; Daly, M. E. op. cit., 73.

21 Ó Gráda, C. (1994) *Ireland; A New Economic History, 1780-1939*, 304; Bielenberg, A. (1991) *Cork's Industrial Revolution 1780-1880*, 52 and 67; Bell, J. and Watson, M. (1986) *Irish Farming - Implements and Techniques 1750-1900*, 180; Gillmor, D. A. (1977) *Agriculture in the Republic of Ireland*, 81-3.

References

22 Evans, E. E. (1942) *Irish Heritage*, 128; Crawford, E. M. (1981) op. cit., 117; O'Sullivan, D. J. op. cit., 23 and 35; Gillmor, D. A. op. cit., 79-81.

23 Bell and Watson, op. cit., 179-180; Glassie, H. (1982) *Passing the Time*, 416; Evans, E. E. (1957) op. cit., 201-211; Evans, E. E. (1942) op. cit., 128-9; Buchanan, R. H. (1988) 'Calender Customs', 56-7 and 64.

BIBLIOGRAPHY

Bamfords' General Catalogue of Agricultural Machines, No. 27, November, 1894.

Barber, J. and Deady, J. (1970) 'A Saddle-Quern from Donickmore, Co. Cork', *Journal of the Cork Historical and Archaeological Society*, 75, 221: 54-55.

Bell, J. (1981) 'Sickles, Hooks, and Scythes in Ireland', *Folk Life*, 19: 26-35.

Bell, J. (1982-83) 'Recent Evidence for the Use of Ard Ploughs in County Fermanagh', *Sinsear*: 30-34.

Bell, J. (1983a) 'The Use of Oxen on Irish Farms since the Eighteenth Century', *Ulster Folklife*, 29: 18-28.

Bell, J. (1983b) 'Harrows Used in Ireland', *Tools & Tillage*, 4, 4: 195-205.

Bell, J. (1984) 'A Contribution to the Study of Cultivation Ridges in Ireland', *Journal of the Royal Society of Antiquaries of Ireland*, 114: 80-97.

Bell, J. and Watson, M. (1986) *Irish Farming - Implements and Techniques 1750-1900*, Edinburgh: John Donald Publishers Ltd.

Bielenberg, A. (1991) *Cork's Industrial Revolution 1780-1880. Development or Decline?*, Cork: Cork University Press.

Blandford, P. W. (1976) *Old Farm Tools and Machinery - An Illustrated History*, Canada: Douglas David & Charles Ltd.

Bourke, P. M. A. (1976) ' The Irish grain trade, 1839-48', *Irish Historical Studies*, 20, 78: 156-169.

Bowie, G. (1979) 'Corn Drying Kilns, Meal Milling and Flour in Ireland, *Folk Life*, 17: 5-13.

Buchanan, R. A. (1982) *Industrial Archaeology in Britain*, Penguin Books Ltd.: England.

Buchanan, R. H. (1988) 'Calender Customs', *Seasonal Customs*, 10, 2: 33-71.

Clarkson, L. A. and Crawford, E. M. (1988) 'Dietary Directions: A Topographical Survey of Irish Diet, 1836', in Mitchison, R. and Roebuck, P. (eds.) *Economy and Society in Scotland and Ireland 1500-1939*, Edinburgh: John Donald Publishers Ltd: 171-192.

Collins, E. J. T. (1972) 'The Diffusion of the Threshing Machine in Britain, 1790-1880', *Tools & Tillage*, 2, 1: 16-33.

Conboy, N. (1982-83) 'Burying the Sheaf - A form of Murder by Magic', *Sinsear*: 102-106.
Crawford, E. M. (1981) 'Indian Meal and Pellagra in Nineteenth-Century Ireland', in Goldstrom, J. M. and Clarkson, L. A. (eds.) *Irish Population, Economy, and Society - Essays in Honour of the late K. H. Connell*, Oxford: Clarendon Press/Oxford University Press: 113-133.
Crotty, R. D. (1966) *Irish Agricultural Production - Its Volume and Structure*, Cork: Cork University Press.
Crowley, F. (1979) *In West Cork Long Ago*, Cork and Dublin: The Mercier Press.
Cullen, L. M. (1977) 'Eighteenth-Century Flour Milling in Ireland', *Irish Economic and Social History*, 4: 5-25.
Cullen, L. M. (1981a) 'Population Growth and Diet, 1600-1850', in Goldstrom, J. M. and Clarkson, L. A. (eds.) *Irish Population, Economy, and Society - Essays in Honour of the late K. H. Connell*, Oxford: Clarendon Press/Oxford University Press: 89-112.
Cullen, L. M. (1981b) *The Emergence of Modern Ireland 1600-1900*, New York: Holmes & Meier Publishers Inc.
Cullen, L. M. (1983) 'Incomes, Social Classes and Economic Growth in Ireland and Scotland, 1600-1900', in Devine, T. M. and Dickson, D. (eds.) *Ireland and Scotland 1600-1850*, Edinburgh: John Donald Publishers Ltd.: 248-260.
Cullen, L. M. (1987) *An Economic History of Ireland since 1660*, London: B.T. Batsford Ltd.
Daly, M. E. (1981) *Social and Economic History of Ireland since 1800*, Dublin: The Educational Company of Ireland Ltd.
Danaher, K. (1962) *In Ireland Long Ago*, Dublin and Cork: The Mercier Press.
Danaher, K. (1972) *The Year in Ireland*, Cork: The Mercier Press.
Danaher, K. (1985) *The hearth and stool and all! Irish Rural Households*, Cork and Dublin: The Mercier Press.
Department of Agriculture and Technical Instruction (1902) *Ireland - Industrial and Agricultural*, Dublin, Cork, Belfast: Browne & Nolan Ltd.
Dickson, D. (1983) 'The Place of Dublin in the Eighteenth-Century Irish Economy', in Devine, T. M. and Dickson, D. (eds.) *Ireland*

and Scotland 1600 - 1850, Edinburgh: John Donald Publishers Ltd.: 177-192.

Dickson, D. (1987) *New Foundations: Ireland 1660-1800*, Dublin: Helicon Ltd.

Edwards, N. (1990) *The Archaeology of Early Medieval Ireland*, London: B. T. Batsford Ltd.

Evans, E. E. (1942) *Irish Heritage - The Landscape, The People and their Work*, Dundalk: W. Tempest, Dundalgan Press.

Evans, E. E. (1957) *Irish Folk Ways*, London and Boston: Routledge & Kegan Paul Ltd.

Evans, E. E. (1981) *The Personality of Ireland - Habitat, Heritage and History*, Belfast: Blackstaff Press Ltd.

Flanagan, L. (1992) *A Dictionary of Irish Archaeology*, Dublin: Gill and Macmillan Ltd.

Gahan, W. (1956) 'The Origins of Crop Husbandry', *Biatas -The Beat Grower*, 10, 6: 249-256.

Gailey, A. (1970) 'The Typology of the Irish Spade', in Gailey, A. and Fenton, A. (eds.) *The Spade In Northern and Atlantic Europe*, Co. Down: Ulster Folk Museum and Institute of Irish Studies, Queen's University Belfast: 35-48.

Gailey, A. (1971) 'Spade Tillage in South-West Ulster and North Connacht', *Tools & Tillage*, 1, 4: 225-236.

Gailey, A. (1972) 'A family spade-making business in county Tyrone', *Folk Life*, 10: 26-46.

Gailey, A. (1984) 'Introduction and Spread of the Horse-powered Threshing Machine to Ulster's Farms in the Nineteenth Century: some aspects', *Ulster Folklife*, 30: 37-54.

Gailey, A. (1990) '"...such as pass by us daily...": The Study of Folklife', *Ulster Folklife*, 36: 4-22.

Gillespie, R. (1984) 'Harvest Crises in Early Seventeenth-Century Ireland', *Irish Economic and Social History*, 11: 5-18.

Gillmor, D. A. (1977) *Agriculture in the Republic of Ireland*, Budapest: Akadémiai Kiadó.

Glassie, H. (1982) *Passing the Time - Folklore and History of an Ulster Community*, Dublin: The O'Brien Press Ltd.

Graham, B. J. (1993) 'The High Middle Ages: *c.* 1100 to *c.* 1350', in Graham, B. J. and Proudfoot, L. J. *An Historical Geography of Ireland*, London: Academic Press Ltd: 58-98.

Green, E. R. R. (1969) 'Industrial Decline in the Nineteenth Century', in Cullen, L. M. (ed.) *The Formation of the Irish*

Economy, Cork: The Mercier Press: 89-100.
Hamlin, A. (1981) 'Using Mills on Sunday', in Scott B.G. (ed.) *Studies on Early Ireland: Essays in Honour of M. V. Duignan*, Belfast: Association of Young Irish Archaeologists: 11.
Holt, R. (1990) 'Milling Technology in the Middle Ages: the Direction of Recent Research', *Industrial Archaeology Review*, 13, 1: 50-58.
Illustrated London News, 19/7/1851 and 15/11/1862.
International Exhibition of 1862: The Illustrated Catalogue of the Industrial Department British Division, Vol. 1.
Irish Penny Journal, 6/2/1841, Vol. 1
Johnson, J. H. (1967) 'Harvest Migration from Nineteenth-Century Ireland', *Transactions, Institute of British Geographers*, 41: 97-112.
Kennedy, L. (1981) 'Regional Specialization, Railway Development, and Irish Agriculture in the Nineteenth Century', in Goldstrom, J. M. and Clarkson, L. A. (eds.) *Irish Population, Economy, and Society - Essays in Honour of the late K. H. Connell*, Oxford: Clarendon Press/Oxford University Press: 173-193.
Lucas, A. T. (1972-75) 'Irish Ploughing Practices', *Tools & Tillage*, 2, 1: 52-62 (Part 1); 2, 2: 67-83 (Part 2); 2, 3: 149-160 (Part 3); 2, 4: 195-210 (Part 4).
Lucas, A. T. (1978) 'The "Gowl-Gob" - An Extinct Spade Type from County Mayo, Ireland', *Tools & Tillage*, 3, 3: 191-199.
McCutcheon, W. A. (1980) *The Industrial Archaeology of Northern Ireland*, Rutherford, Madison, Teaneck: Fairleigh Dickinson University Press.
MacEoin, G. (1981) 'The Early Irish Vocabulary of Mills and Milling', in Scott B.G. (ed.) *Studies on Early Ireland: Essays in Honour of M. V. Duignan*, Belfast: Association of Young Irish Archaeologists: 13-19.
MacNeill, M. (1982) *The Festival of Lughnasa - A Study of the Survival of the Celtic Festival of the Beginning of Harvest*, Dublin: Comhairle Bhéaloideas Éireann.
Mahon, B. (1980) 'Oatmeal as Food' *Sinsear* , 2: 123-127.
Mitchell, F. (1986) 'Early Settlement and Society before 500 AD', in Nolan, W. (ed.) *The Shaping of Ireland - The Geographical Perspective*, Cork and Dublin: The Mercier Press/RTE: 28-43.

Mokyr, J. (1981) 'Irish History with the Potato', *Irish Economic and Social History*, 8: 8-29.

Monk, M. A. (1981) 'Post-Roman Drying Kilns and the Problem of Function: a preliminary statement', in Ó Corráin, D. (ed.) *Irish Antiquity - Essays and Studies presented to Professor M. J. O'Kelly*, Cork: Tower Books: 216-230.

Neill-Watson, J. (1993) *A History of Farm Mechanisation in Ireland 1890-1990*, Dublin: Farm Tractor and Machinery Trade Association.

Ó Corráin, D. (1972) *Ireland before the Normans*, Dublin: Gill and Macmillan Ltd.

Ó Danachair, C. (1955) 'The Flail and other Threshing Methods', *Journal of the Cork Historical and Archaeological Society*, 60, 191: 6-14.

Ó Danachair, C. (1963) 'The Spade in Ireland', *Béaloideas*, 31: 98-114.

Ó Danachair, C. (1970a) 'The Use of the Spade in Ireland', in Gailey, A. and Fenton, A. (eds.) *The Spade In Northern and Atlantic Europe*, Co. Down: Ulster Folk Museum and Institute of Irish Studies, Queen's University Belfast: 49-56.

Ó Danachair, C. (1970b) 'The Flail in Ireland', *Ethnologia Europaea*, 4: 50-55.

Ó Gráda, C. (1988) *Ireland before and after the Famine - Explorations in economic history, 1800-1925*, Manchester: Manchester University Press.

Ó Gráda, C. (1994) *Ireland - A New Economic History 1780-1939*, Oxford: Clarendon Press/Oxford University Press.

O'Dowd, A. (1981) *Meitheal - A Study of Co-operative Labour in Rural Ireland*, Dublin: Comhairle Bhéaloideas Éireann.

O'Mahony, S. (1993) *A Gate to the Past - The History and Folklore of Carrigaline*, Vol. 1, Carrigaline: Shannon Park Press.

O'Neill, T. P. (1973) 'Poverty in Ireland 1815-45', *Folk Life*, 11: 22-33.

O'Neill, T. P. (1977) *Life and Tradition in Rural Ireland*, London: J. M. Dent & Sons Ltd.

O'Neill, T. P. (Undated) 'Tools and Things: Machinery on Irish Farms 1700-1981', in Gailey, A. and Ó hÓgáin, D. *Gold Under the Furze: Studies in Folk Tradition*, Dublin: The Glendale Press: 101-114.

O'Sullivan, A. M. (Undated) *Tacumshin Windmill - its history and mode of operation.*
O'Sullivan, D. J. (1995) *The Story of Corn and Flour Milling in Ireland,* Tralee: Kingdom Printers.
O'Sullivan, J. C. (1972) 'Iron-Shod Wooden Spade from Ballynakillew Mountain, Co. Donegal', *Journal of the Royal Society of Antiquaries of Ireland,* 102: 244-246.
Paterson, T. G. F. (1944) 'Harvest Customs in Co. Armagh', *Ulster Journal of Archaeology,* 7: 109-116.
Perren, R. (1990) 'Structural change and market growth in the food industry: flour milling in Britain, Europe, and America, 1850-1914', *Economic History Review,* second series, 43, 3: 420-437.
Pollack, V. (1988) 'Work On the Land', *Seasonal Customs,* 10, 2: 14-21.
Power, D. et al (1992) *Archaeological Inventory of County Cork,* (Vol. 1: West Cork), Dublin: Stationery Office.
Raftery, B. (1984) *La Tene in Ireland - Problems of Origin and Chronology,* Marburg: Vorgeschichtliches Seminars.
Ransomes, Sims & Jefferies' Illustrated Catalogue of Steam Thrashing Machinery, July, 1921.
Rynne, C. (1989a) 'Archaeology and The Early Irish Watermill', *Archaeology Ireland,* 3, 3: 110-114.
Rynne, C. (1989b) 'The Introduction of the Vertical Watermill into Ireland: Some Recent Archaeological Evidence', *Medieval Archaeology,* 33: 21-31.
Rynne, C. (1990) 'Some Observations on the Production of Flour and Meal in the Early Historic Period', *Journal of the Cork Historical and Archaeological Society,* 95, 254: 20-29.
Rynne, C. (1992) 'Milling in the 7th century - Europe's Earliest Tide Mills', *Archaeology Ireland,* 6, 2: 22-24.
Rynne, C. (1993) *The Archaeology of Cork City and Harbour from the Earliest Times to Industrialisation,* Cork: The Collins Press.
Shaffrey, P. and M. (1985) *Irish Countryside Buildings - Everyday Architecture in the Rural Landscape,* Dublin: The O'Brien Press.
Sharkey, O. (1985) *Old Days, Old Ways - An Illustrated Folk History of Ireland,* Dublin: The O'Brien Press Ltd.
Siguat, F. (1988) 'A Method for Identifying Grain Storage Techniques and its Application for European Agricultural History', *Tools & Tillage,* 6, 1: 3-32.

Simms, A. (1986) 'Continuity and Change: Settlement and Society in Medieval Ireland *c.* 500-1500', in Nolan, W. (ed.) *The Shaping of Ireland - The Geographical Perspective*, Cork and Dublin: The Mercier Press/RTE: 44-65.

Smyth, W. J. (1993) 'The Making of Ireland: Agendas and Perspectives in Cultural Geography', in Graham, B. J. and Proudfoot, L. J. *An Historical Geography of Ireland*, London: Academic Press Ltd.: 399-438.

Sproule, J. (ed.) (1854) *The Irish Industrial Exhibition of 1853*, Dublin.

Tighe, W. (1802) *Statistical Observations relative to the County of Kilkenny*, Dublin.

Tucker, D. G. (1987) 'Millstone making in England', *Industrial Archaeology Review*, 9, 2: 167-188.

Turner, M. (1987) 'Towards an Agricultural Prices Index for Ireland 1850-1914', *The Economic and Social Review*, 18, 2: 123-136.

Turner, M. (1993) 'Rural Economies in Post-Famine Ireland, *c.* 1850-1914', in Graham, B. J. and Proudfoot, L. J. *An Historical Geography of Ireland*, London: Academic Press Ltd.: 293-337.

Watson, M. (1982) 'North Antrim Swing Ploughs: their Construction and Use', *Ulster Folklife*, 28: 13-23.

Watson, M. (1985) 'Common Irish Plough Types and Tillage Techniques', *Tools and Tillage*, 5, 2: 85-98.

Whelan, K. (1986) 'The Famine and Post-Famine Adjustment', in Nolan, W. (ed.) *The Shaping of Ireland - The Geographical Perspective*, Cork and Dublin: The Mercier Press/RTE: 151- 164.

FURTHER READING

Ballard, L. M. (1988) 'When Does the Sun Dance?', *Seasonal Customs*, 10, 2: 5-9.

Bell, J. (1978) 'Implements on Irish Farms in 1900', *Ulster Folk & Transport Museum Year Book 1976/77*: 18-20.

Bell, J. (1978) 'Relations of Mutual Help between Ulster Farmers', *Ulster Folklife*, 24: 48-58.

Bell, J. (1980) 'Wooden Ploughs from the Mountains of Mourne, Ireland', *Tools & Tillage*, 4, 1: 46-57.

Bell, J. (1985) 'Last Sheaves, Ancient Cattle and Protestant Bibles', *Béaloideas*, 53: 5-10.

Bell, J. and Watson, M. (1988) *Farming in Uster - Historic photographs of Ulster farming and food*, Belfast: The Friar's Bush Press.

Berg, G. (1976) 'The Introduction of the Winnowing-machine in Europe in the 18th Century', *Tools & Tillage*, 3, 1: 25-46.

Bourke, P. M. A. (1965) 'The Agricultural Statistics of the 1841 Census of Ireland. A Critical Review', *Economic History Review*, second series, 18, 2: 376-391.

Bowie, G. (1976) 'Watermills, Windmills and Stationery Steam Engines in Ireland', (Ph.D. thesis abstract), *Irish Economic and Social History*, 3: 80-82.

Buchanan, R. H. (1965) 'Tradition and Change in Rural Ulster', *Folk Life*, 3: 39-45.

Campbell, A. (1935) 'Irish Fields and Houses - A Study of Rural Culture', *Béaloideas*, 5, 1: 57-74.

Cheape, H. (1978) 'Questionnaires and the Grain Harvest in Scotland', *Folk Life*, 16: 27-41.

Collins, E. J. T. (1987) 'The Rationality of "Surplus" Agricultural Labour: Mechanization in English Agriculture in the Nineteenth Century', *Agricultural History Review*, 35, Part 1: 36-46.

Colyer, R. J. (1982-83) 'Crop Husbandry in Wales Before the Onset of Mechanization', *Folk Life*, 21: 49-70.

Crawford, W. H. and Gailey, A. (1981) 'Ulster and Industrial Archaeology', *Ulster Folklife*, 27: 81-87.

Danaher, K. (1970) *The Pleasant Land of Ireland*, Cork: The Mercier Press.

Donnelly, J. S. (1975) *The Land and the People of Nineteenth-Century Cork - The Rural Economy and the Land Question*, London and Boston: Routledge & Kegan Paul Ltd.

Donnelly, J. S. (1976) 'The Irish Agricultural Depression of 1859-64', *Irish Economic and Social History*, 3: 33-54.

Doyle, J. (1985) 'The Water Mills of Kilkenny', *Old Kilkenny Review*, 3, 2: 147-160.

Fahy, E. M. (1956) 'A Horizontal Mill at Mashanaglass, Co. Cork', *Journal of the Cork Historical and Archaeological Society*, 61, 193: 13-57.

Fenton, A. (1974) 'The Cas-Chrom: A Review of the Scottish Evidence', *Tools & Tillage*, 2, 3: 131-148.

Gailey, A. (1964) 'The Ulster Tradition', *Folk Life*, 2: 27-41.

Gailey, A. (1966) 'The Disappearance of the Horse from the Ulster Farm, *Folk Life*, 4: 51-55.
Gailey, A. (1978) *Ulster Folk Ways - An Introduction*, The Irish Heritage Series, 20, Belfast: Eason & Son (N.I.) Ltd.
Gosling, P., Duffy, P. and Delaney, D. (1992) 'Thoor Ballylee - place of parapets, poems and mill wheels', *Archaeology Ireland*, 6, 3: 25-27.
Jones, G. E. (1965) 'The nature and consequences of technical change in farming', *Folk Life*, 3: 79-87.
Lucas, A. T. (1953) 'The Horizontal Mill in Ireland', *Journal of the Royal Society of Antiquaries of Ireland*, 83: 1-36.
Lucas, A. T. (1955) 'A Horizontal Mill at Ballykilleen, Co. Offaly', *Journal of the Royal Society of Antiquaries of Ireland*, 85: 100-113.
Mokyr, J. (1980) ' Industrialization and Poverty in Ireland and the Netherlands', *Journal of Interdisciplinary History*, 10, 3: 429-458.
Mokyr, J. (1983) *Why Ireland Starved: A Quantitative and Analytical History of the Irish Economy, 1800-1850*, London: George Allen & Unwin.
Monk, M. (1987) 'Experimental Archaeology', *Archaeology Ireland*, 1, 2: 65-67.
Newman, P. R. (1985-86) 'The Flail, the Harvest and Rural Life', *Folk Life*, 24: 70-86.
North, D. J. (1984-85) 'The One-Way Plough in South-Eastern England', *Folk Life*, 23: 49-64.
Note (1985) 'Illustrations of the Use of Oxen in Ireland', *Ulster Folklife*, 31: 73-75.
Ó Danachair, C. (1957) 'Some Distribution Patterns in Irish Folk Life', *Béaloideas*, 25: 108-123.
Ó Danachair, C. (1965) 'Distribution Patterns in Irish Folk Tradition', *Béaloideas*, 33: 97-113.
Ó Danachair, C. (ed.) (1976) *Folk & Farm - Essays in Honour of A. T. Lucas*, Dublin: Royal Society of Antiquaries of Ireland.
Ó Gráda, C. (1975) 'Supply Responsiveness in Irish Agriculture during the Nineteenth Century', *Economic History Review*, second series, 28, 2: 312-317.
Ó Gráda, C. (1989) *The Great Irish Famine*, Dublin: Gill & Macmillan Ltd.

O'Brien, J. B. (1977) 'Agricultural Prices and Living Costs in Pre-Famine Cork', *Journal of the Cork Historical and Archaeological Society*, 82, 235: 1-10.

O'Connor, R. and Guiomard, C. (1985) 'Agricultural Output in the Irish Free State Area before and after Independence', *Irish Economic and Social History*, 12: 89-97.

O'Curry, E. (1971) *On the Manners and Customs of the Ancient Irish*, New York: Lemma Publishing Corp. (Originally published, 1873).

O'Dowd, A. (1981) 'The Miller and his Tale', *Sinsear*, 3: 83-91.

O'Flanagan, P., Ferguson, P. and Whelan, K. (eds.) (1987) *Rural Ireland 1600-1900: Modernisation and Change*, Cork: Cork University Press.

O'Flanagan, P. and Buttimer, C. G. (eds.) (1993) *Cork - History and Society: Interdisciplinary Essays on the History of an Irish County*, Dublin: Geography Publications.

Paterson, T. G. F. (1942) 'Harvest Customs in County Armagh', *Ulster Journal of Archaeology*, 5: 2-7.

Perkins, J. A. (1976-77) 'Harvest Technology and Labour Supply in Lincolnshire and the East Riding of Yorkshire 1750-1850', *Tools & Tillage*, 3, 1: 47-64; 3, 2: 125-135.

Perkins, J. A. (1977-78) 'The Sickle, the Scythe, and the Physical Characteristics of Migratory Harvest Workers', *Ethnologia Europaea*, 10, 1: 88-97.

Power, D. (1989) 'County Focus: Cork', *Archaoelogy Ireland*, 3, 2: 46-50.

Robinson, P. (1982) 'A Water-Mill Built in 1615 by the Drapers' Company at Moneymore, County Londonderry', *Ulster Folklife*, 28: 49-55.

Rynne, C. (1991) 'Cork's Industrial Past', *Technology Ireland*, 22, 9: 36-38.

Shaffrey, P. and M. (1983) *Buildings of Irish Towns - Treasures of Everyday Architecture*, Dublin: The O'Brien Press.

Siguat, F. (1989) 'Storage and Threshing in Preindustrial Europe: Additional Notes', *Tools & Tillage*, 6, 2: 119-124.

Solar, P. (1985) 'The Reconstruction of Irish External Trade Statistics for the Nineteenth Century', *Irish Economic and Social History*, 12: 63-78.

Thompson, A. (1977) 'The Origins of a Harvesting Revolution: The Development of the Combine Harvester in Australia, Canada and U.S.A., *Tools & Tillage*. 3, 2: 67-77.
Turner, M. (1990) 'Output and Productivity in Irish Agriculture from the Famine to the Great War', *Irish Economic and Social History*, 17: 62-78.
Vaughan, W. E. (1990) 'Potatoes and Agricultural Output', *Irish Economic and Social History*, 17: 79-92.
Williams, F. (1989-90) 'An Irish Threshing Proverb, *Folk Life*, 28: 41-45.
Zanden, J. L. van (1991) 'The first green revolution: the growth of production and productivity in European agriculture, 1870-1914, *Economic History Review*, 44, 2: 215-239.

INDEX